# Environmental Sustainability in the Food Industry

Criticism facing the food processing industry includes adverse ecological impacts like decline in biodiversity, environmental degradation, water pollution, eutrophication, greenhouse gas emissions, and the loss of agricultural land. ***Environmental Sustainability in the Food Industry: A Green Perspective*** delves into the effects of food processing on the environment, human health, nutrition, energy efficiency, nanotechnology in the food industry, and the maintenance of ecological sustainability.

This book presents eco-friendly approaches to reducing the impacts of food processing on the environment and to promoting sustainable development. The focus of this text is how to implement green practices in the food industry in order to reduce the negative impacts of food processing on the ecosystem, as well as to improve food quality for better human health and nutrition. The text also explains the food industry's focus on sustainable aspects in resource conservation and reduction of energy consumption.

**Key Features:**

- Describes the contributions of the food industry sector on human health and nutrition
- Covers eco-friendly approaches to reducing the negative impacts of food processing on the environment
- Discusses the uses of advanced techniques such as nanotechnology, non-thermal techniques, and more to improve food processing

***Environmental Sustainability in the Food Industry*** highlights the food industry and environmental issues. It is a great resource for students, researchers, and professionals alike, as well as anyone with an interest in green paths to food quality and nutrition.

# Environmental Sustainability in the Food Industry

A Green Perspective

Edited by
Bharat Kapoor, Rahul Singh,
Dhriti Kapoor, and Vandana Gautam

CRC Press
Taylor & Francis Group
Boca Raton London New York

First edition published 2023
by CRC Press
6000 Broken Sound Parkway NW, Suite 300, Boca Raton, FL 33487-2742

and by CRC Press
4 Park Square, Milton Park, Abingdon, Oxon, OX14 4RN

*CRC Press is an imprint of Taylor & Francis Group, LLC*

© 2023 selection and editorial matter, Bharat Kapoor, Rahul Singh, Dhriti Kapoor and Vandana Gautam; individual chapters, the contributors

Reasonable efforts have been made to publish reliable data and information, but the author and publisher cannot assume responsibility for the validity of all materials or the consequences of their use. The authors and publishers have attempted to trace the copyright holders of all material reproduced in this publication and apologize to copyright holders if permission to publish in this form has not been obtained. If any copyright material has not been acknowledged please write and let us know so we may rectify in any future reprint.

Except as permitted under U.S. Copyright Law, no part of this book may be reprinted, reproduced, transmitted, or utilized in any form by any electronic, mechanical, or other means, now known or hereafter invented, including photocopying, microfilming, and recording, or in any information storage or retrieval system, without written permission from the publishers.

For permission to photocopy or use material electronically from this work, access www.copyright.com or contact the Copyright Clearance Center, Inc. (CCC), 222 Rosewood Drive, Danvers, MA 01923, 978-750-8400. For works that are not available on CCC please contact mpkbookspermissions@tandf.co.uk

*Trademark notice*: Product or corporate names may be trademarks or registered trademarks and are used only for identification and explanation without intent to infringe.

*Library of Congress Cataloging-in-Publication Data*
Names: Kapoor, Bharat, editor. | Singh, Rahul (Biologist), editor. |
Kapoor, Dhriti, editor. | Gautam, Vandana, editor.
Title: Environmental sustainability in the food industry : a green perspective /
edited by Bharat Kapoor, Rahul Singh, Dhriti Kapoor, and Vandana Gautam.
Description: First edition. | Boca Raton, FL : CRC Press, 2023. |
Includes bibliographical references and index.
Identifiers: LCCN 2022025658 (print) | LCCN 2022025659 (ebook) |
ISBN 9781032193038 (hbk) | ISBN 9781032164618 (pbk) | ISBN 9781003258568 (ebk)
Subjects: LCSH: Food industry and trade–Environmental aspects. |
Food science–Safety measures. | Sustainable engineering.
Classification: LCC TD195.F57 E547 2023 (print) |
LCC TD195.F57 (ebook) | DDC 664.0028/6–dc23/eng/20221011
LC record available at https://lccn.loc.gov/2022025658
LC ebook record available at https://lccn.loc.gov/2022025659

ISBN: 9781032193038 (hbk)
ISBN: 9781032164618 (pbk)
ISBN: 9781003258568 (ebk)

DOI: 10.1201/9781003258568

Typeset in Times
by Newgen Publishing UK

# Contents

Preface ....................................................................................................................vii
About the Editors ....................................................................................................ix
List of Contributors ................................................................................................xi

Chapter 1
Novel and Innovative Strategies for Food Packaging Processes ............................ 1

**Sadaf Jan, Savita Bhardwaj, Tunisha Verma, Renu Bhardwaj, Dhriti Kapoor, and Rattandeep Singh**

Chapter 2
Contaminants and Their Control Measures to Maintain Hygiene in Food Processing ................................................................................................ 17

**Varsha Singh and Amit Wadehra**

Chapter 3
Green Practices Initiative as a Sustainable Aspect for Food Industry ................... 33

**Deepika Puri and Bharat Kapoor**

Chapter 4
Improving Food Processing Using Various Technologies in the Food Industry ......................................................................................................... 45

**Monika Thakur and Sayeeda Kousar Bhatti**

Chapter 5
Functional Properties of Food Processing as a Novel Technology for Human Health and Nutrition ........................................................................... 59

**Ruby Angurana, Vaidehi Katoch, Tunisha Verma, and Savita Bhardwaj**

Chapter 6
Food Processing Potential for Energy Efficiency and Use .................................... 79

**Dhriti Sharma, Savita Bhardwaj, Tunisha Verma, Mamta Pujari, Rahul Singh, and Vandana Gautam**

Chapter 7
Global Food Security and Effects of Various Environmental Constraints on Food Crops ..................................................................................... 95

**Tunisha Verma, Sahima Tabasum, Savita Bhardwaj, Vandana Gautam, Bharat Kapoor, and Dhriti Kapoor**

Chapter 8
Food Safety, Quality, and Policies ......................................................................... 109
**Ajay Rathore and Ambika Bhatia**

Chapter 9
Emerging Role of Nanotechnology in the Food Industry for Food
Processing and Packaging ..................................................................................... 121
**Nitika Kapoor**

Index ....................................................................................................................... 145

# Preface

Food is a vital part of our lives, and food-related industries are the basic and important requirements of the entire world's population due to the growing interest of consumers in knowing more about food quality and nutritional worth. Development in the food industry sector has increased in the past few decades because of the increase in the need for safe and healthy food. This book is mainly about the effect of food processing on different aspects such as the environment, human health, nutrition, energy efficiency, nanotechnology in the food industry, and maintaining ecological sustainability. Numerous challenges have been faced by the food industry such as criticism, climate change, environmental contamination, alterations in soil and water properties, deforestation, and more, but the food industry puts greater efforts into disreputing its critics by making a sustainable food system. The food industry sector appears as a vital moiety which provides food for human consumption and fulfills customer needs by providing food with a high nutritional content. The food industry comprises several subject areas, i.e., agriculture, food processing, food distribution, regulation, financial services, research and development, and marketing, among which food processing acts as one of the most important areas because it helps to improve food characteristics such as flavor, texture, taste, shelf life, appearance, and more. There are a number of food processing methods practiced individually or as an amalgamation of other methods, such as extraction, partition, manufacture, physical progressions, purification, heat methods, biochemical progressions, and more. All of these are categorized into thermal and non-thermal processing methods.

Demand for innovative, new food packaging approaches has increased due to customer need for good quality, easy preparation, quick availability, tastiness, moderately processed food moieties and recent changes in lifestyle. Obstructed respiratory routes, inhibition of oxidation and microbial attacks, and sustained release of antioxidants during storage are the major changes caused by the improvement in new food packaging techniques. The food industry is also focusing on sustainable ways for conservation of energy and reduction of energy consumption, particularly due to the escalation in the demand for food safety. Therefore, there is a drastic improvement in the ecosystem footprint of the food industry.

Food quality and nutrition is highly impacted by the excessive practices of fertilizers, pesticides, and biological or chemical compounds. To maintain overall food quality, packaging materials are vital for the protection of food products from the various contaminants. In this context, nanotechnology has gained increased attention from researchers in the food industry, particularly in food processing, packaging, storage, and manufacture of innovative products. Another major goal of the food industry is to boost consumer health. The agronomic sector enhances crop yield via development of novel and innovative techniques. Hence, the food processing industry needs to be agile if it is going to adapt new methods to further improve the quality of these crop products. Therefore, food industries are adopting several techniques, which can increase human health and nutrition. Overall, this book will describe aspects related to food industry and environmental issues.

## About the Editors

**Bharat Kapoor,** MHM, MBA (Hospitality Management), PhD, is assistant professor in the Department of Hotel Management and Tourism at Guru Nanak Dev University, Amritsar, Punjab, India. Dr. Kapoor has 19 years of academic and industrial experience in the field of Hospitality. During his industrial experience he worked with Taj Group of Hotels, The Imperial New Delhi, and the ITC Group of hotels in the departments of food and beverages. As far as his academic experience is concerned, he held the position of HOD–Hotel Management at Lovely Professional University Jalandhar, Director of Hospitality at CT University, Ludhiana, and Dean of the Doctoral Research Centre at Chitkara University, Chandigarh, India. He earned his PhD in Hotel Management from Kurukshetra University, Haryana in 2012. He has had more than 25 publications in reputed journals, more than 10 book chapters, two patents granted, and two International books. He has organized many National and International conferences. He is a member of various professional bodies related to the fields of Hospitality and Tourism and he is also a member of the board of studies of various reputed universities in India.

**Rahul Singh** is associate professor in the Department of Zoology at Lovely Professional University, Phagwara, Punjab, India. Dr. Singh has 10 years of teaching and research experience in the field of zoology. He has earned his PhD from Dr. Ram Manohar Lohia, Avadh University, Faizabad, U.P., India. Currently, his research focus is on aquatic emerging pollutants and maintenance of fish health with probiotics. He has published 16 research papers in reputed journals and two book chapters.

**Dhriti Kapoor** is assistant professor in the Department of Botany at Lovely Professional University, Phagwara, Punjab, India. She completed her postgraduate study in botany in 2009, master's of philosophy (MPhil) in 2011, and doctorate in 2015 at Guru Nanak Dev University, Amritsar, Punjab. Her research areas are plant stress physiology, plant biochemistry, phytoremediation, plant growth regulation, and ecotoxicology. She has published more than 50 research papers in peer-reviewed journals and 42 book chapters in internationally published volumes such as Springer, Elsevier, Nova, and Wiley. At present, she is engaged in studying the physiological, biochemical, and molecular aspects of different agricultural crops under biotic as well as abiotic stress conditions.

**Vandana Gautam** has been teaching as part of the faculty of the College of Horticulture and Forestry (Dr. Y. S. Parmar University of Horticulture and Forestry, Nauni, India) since 2017. She earned her postgraduate (MSc), master's of philosophy (MPhil), as well as doctorate (PhD) degrees in the subject of environmental sciences from Guru Nanak Dev University, Amritsar, Punjab, India. Her main research area is phytoremediation and environmental stress physiology. She has published more than 20 research papers in peer-reviewed journals and 20 book chapters in internationally published volumes such as Springer, Nova, Elsevier, and Wiley. She is a recipient of the Junior Research Fellowship and Senior Research Fellowship by UGC, New Delhi, India. Dr. Gautam is dynamically engaged in studying the physiological, biochemical, and molecular responses of different horticultural and agricultural plants under environmental stress.

# Contributors

**Ruby Angurana**
Department of Zoology
School of Bioengineering and
 Biosciences
Lovely Professional University
Phagwara (Punjab), India

**Renu Bhardwaj**
Department of Botanical and
 Environmental Sciences
Guru Nanak Dev University
Amritsar (Punjab), India

**Savita Bhardwaj**
Department of Botany
School of Bioengineering and
 Biosciences
Lovely Professional University
Phagwara (Punjab), India

**Ambika Bhatia**
Punjabi University
Patiala, India

**Sayeeda Kousar Bhatti**
Department of Botany
Govt Degree College Doda City
Jammu and Kashmir UT, India

**Sadaf Jan**
Department of Biotechnology
School of Bioengineering and
 Biosciences
Lovely Professional University
Phagwara (Punjab), India

**Nitika Kapoor**
PG Department of Botany
Hans Raj Mahila Maha Vidyalaya
Jalandhar (Punjab), India

**Vaidehi Katoch**
Department of Forensic Science
School of Bioengineering and
 Biosciences
Lovely Professional University
Phagwara (Punjab), India

**Mamta Pujari**
Department of Botany
School of Bioengineering and
 Biosciences
Lovely Professional University
Phagwara (Punjab), India

**Deepika Puri**
Chitkara College of Hospitality
 Management
Chitkara University
Punjab, India

**Ajay Rathore**
Punjabi University
Patiala, India

**Dhriti Sharma**
Department of Botany
School of Bioengineering and
 Biosciences
Lovely Professional University
Phagwara (Punjab), India

**Rattandeep Singh**
Department of Biotechnology
School of Bioengineering and
 Biosciences
Lovely Professional University
Phagwara (Punjab), India

**Varsha Singh**
Chitkara University
Punjab, India

**Sahima Tabasum**
Department of Chemistry
School of Chemical Engineering and
  Physical Sciences
Lovely Professional University
Phagwara (Punjab), India

**Monika Thakur**
Division Botany
Department of Bio-Sciences
Career Point University
Hamirpur (H.P.), India

**Tunisha Verma**
Department of Botany
School of Bioengineering and
  Biosciences
Lovely Professional University
Phagwara (Punjab), India

**Amit Wadehra**
Chitkara University
Punjab, India

CHAPTER 1

# Novel and Innovative Strategies for Food Packaging Processes

Sadaf Jan,[1] Savita Bhardwaj,[2] Tunisha Verma,[2]
Renu Bhardwaj,[3] Dhriti Kapoor,[2] and Rattandeep Singh[1*]
[1]Department of Biotechnology, School of Bioengineering and Biosciences,
Lovely Professional University, Phagwara (Punjab), India
[2]Department of Botany, School of Bioengineering and Biosciences,
Lovely Professional University, Phagwara (Punjab), India
[3]Department of Botanical and Environmental Sciences,
Guru Nanak Dev University, Amritsar (Punjab), India
*Corresponding author: rattan.19383@lpu.co.in

## CONTENTS

| | | |
|---|---|---|
| 1.1 | Introduction | 2 |
| 1.2 | Bio-Based Polymers | 3 |
| | 1.2.1 Polymers Sourced from Microbes | 4 |
| | 1.2.2 Polymers Sourced from Wood | 5 |
| | 1.2.3 Polymers Sourced from Proteins | 5 |
| 1.3 | Nanotechnology and Advancement of Barrier Attributes | 6 |
| | 1.3.1 Bio-Nanocomposites and Biodegradable Packaging | 6 |
| 1.4 | Nano-Encapsulation Techniques | 7 |
| 1.5 | Volatile Compounds Utilized in Food Packaging | 7 |
| 1.6 | Fortification of Bio-Plastic Films | 8 |
| | 1.6.1 Bioactive Supplementation with Antioxidant Potential | 8 |
| | 1.6.2 Bio-based Supplementation with Antimicrobial Potential | 9 |
| 1.7 | Conclusion | 9 |
| References | | 10 |

## 1.1 INTRODUCTION

Packaging is imperative for maintaining the quality of foods for use, storage and transportation. Factors like oxygen, water vapor, bacteria, light and other contaminants affect food products which are devoid of packaging after processing (Ahvenainen, 2003). Packaging averts quality deterioration and eases distribution and marketing. Appropriate packaging slows down the deterioration rate thereby augments shelf life of food. The main purpose of food packaging is to preserve, protect and maintain food quality. There is no doubt that contemporary food packaging materials fulfill these requirements. Despite that, the food industry is constantly seeking new innovative technologies to further enhance the quality, shelf life and traceability of their products. The emergence of nanotechnology, which involves the production and utilization of materials ranging in size from 10 to 100 nm has brought new possibilities for the development of novel materials with upgraded properties for use in food packaging (Tsagkaris et al., 2018). With nanotechnology, lightweight packaging can be achieved, possessing much stronger barrier properties and can preserve food from spoiling bacteria. The nano-scale material offers chemical and physical properties in complete contrast to their counterparts because they are a microscopic rather thanmacroscopic material.

The unique features of nanomaterials viz, stronger mechanical strength, substantial surface area-to-volume ratio and distinct optical behavior, mark them preferable for packaging industry. Amalgamating nanomaterials with compatible polymers offers excellent thermal stability, barrier, mechanical and optical properties for packaging compared to those of conventional packaging (Pan and Zhong, 2016; Cerqueira et al., 2018; Eleftheriadou et al., 2017). Plants possess secondary metabolites synthesizing an ability which includes phenols, flavonols, phenolic acids, flavonoids, tannins, flavones, coumarins and quinines. Aforementioned compounds have phenolic structures such as thymol, carvacrol and eugenol, which are highly active against pathogenic organisms. They also manifest anti-microbial properties and provide a defense mechanism against pathogens. In the food industry, anti-microbial material is employed in the form of dips/sprays but this sort of direct use has minimum benefits due to prompt diffusion from the surfaces. Thus immobilizing anti-microbial substances to polymer surfaces gives an anchor to material and thwarts their movement into food thereby maintaining their stability for a longer time period (Ananda et al., 2017).

According to FICCI, the most often used packaging materials are plastics (42%), paper board (31%), metals (15%), glass (7%) and supplementary materials (5%). Plastics are considered to be the most frequently used packaging materials. At present, the global production of plastics is 320 MT per year approximately due to its colossal demand for broad spectrum application (Paletta et al., 2019). For packaging, the commonly exploited polymers are polyethylene terephthalate (PET), polyethylene (PE), polyvinylchloride (PVC), polystyrene (PS), polypropylene (PP), ethylene vinyl alcohol (EVOH) and polyamide (PA) (Luzi et al., 2019). They are chosen for packaging because of their remarkable barrier and mechanical properties, prodigious availability and cost-effectiveness (Park et al., 2017). However, the disposal of plastics as we all know affects the emission of greenhouse gases viz $CO_2$

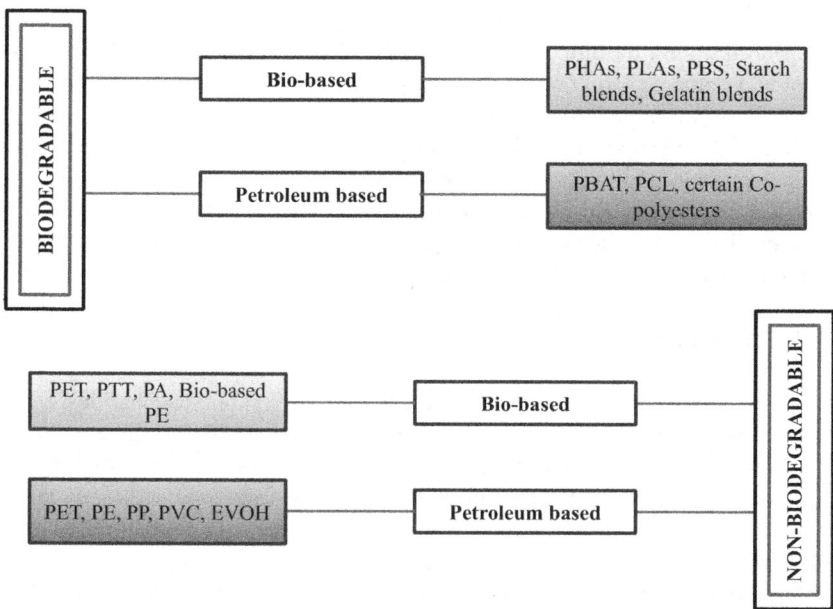

**Figure 1.1** Classification of plastics on the basis of origin and degradability perspective.

and methane leads to environmental concerns (Jain and Tiwari, 2015). Therefore, it is of prime importance to move towards alternative food packaging material. Edible packaging is the thin layer of film or coating cast onto food surfaces that can be consumed as an integral part of a food item. This coating is directly formed onto a food surface whereas film is separately formed and enfolded onto the food surface. Edible materials are only formed when they can produce continuous and cohesive structures. The materials that are edible and proficient at forming continuous and cohesive structures are certified as edible packaging materials (Guilbert et al., 1995). As edible packaging can be devoured along with food and they are produced from organic sources therefore they degrade readily if left unconsumed. By virtue of this, edible packaging has attracted the attention of researchers looking to replace synthetic plastic with this promising alternative (Shit and Shah, 2014). Figure 1.1 illustrates various forms of plastics based on origin and the bio-degradability perspective.

This chapter aims to summarize the findings of novel and innovative technologies used for packaging in the food industry and their progression and potential for future research.

## 1.2 BIO-BASED POLYMERS

Polymers that are produced by living organisms like microbes and plants via metabolic processes are known as bio-polymers. Classification of bio-polymers depends on their origin and can be classified as carbohydrate polymers, for example, cellulose, starch, lignin or chitosan (Hassan et al., 2018; Jiang et al., 2020) proteins, for example, keratin, gelatin or collagen and polyhydroxyalkanoates (Iqbal et al., 2018). Wood fibers (Lignocellulosic) hold a good amount of cellulose and hemicelluloses. The films formed from lignocellulosic fibers give good transparency, tensile strength, better toughness and superior surface gloss (Darni et al., 2017). Cellulose is often modified chemically during the dissolution process to enable the breakdown of polymer chains. The derivatives of cellulose, which are procured after chain fragmentation, can be reformed as coatings.

### 1.2.1 Polymers Sourced from Microbes

The three main polymers based on fermentation are exopolysaccharides, polylactic acid and polyhydroxyalkanoates (Chen, 2010; Alshehrei, 2017; Asghar et al., 2020). Renewable bio-materials are used to produce monomers which are required for the synthesis of bio-poly ethylene, polyhydroxyalkanoates and polylactic acid. Sugarcane and corn are most suitable for manufacturing monomers due to their slight degree of polymerization and ease in extraction. Bio based lactic acid and lactide are produced via fermentation of cornstarch sugars. Regular thermo-plastics need a much higher temperature for processing in comparison with polylactic acid (de Kort et al., 2019). For purifying polylactic acid adsorption, distillation, electrodialysis, reverse osmosis and solvent extraction are needed (Huang and Ramaswamy, 2013). Conversion process expenses can be lowered by high input raw material, but reversibly it will uplift the cost of fermentation and enzyme hydrolysis (Nampoothiri et al., 2010). Vacuum distillation is used for lactide purification. Polymerization and lactic acid condensation is done to acquire polylactic acid (Pal and Katiyar, 2017).

Polyhydroxyalkanoates also are thermally transformed and are produced by renewable raw materials like glucose, fatty acids, maltose through microbial action which is a biotechnology-based conversion (Kawaguchi et al., 2016). Various kinds of polyhydroxyalkanotes have been synthesized such as polyhydroxybutyrate and 3-hydroxybutyrate co 3- hydroxyvalerate. They are used in packaging industries, textile industries and medicinal implants. Due to high costs in production and recovery, polyhydroxyalkanoates are not often used as bioplastics. Exopolysaccharides are produced by different microorganisms such as bacteria, blue green algae and fungi (Asgher et al., 2020). Many types of exopolysaccharides are available including alginate, dextrin, levan, glucans and xanthan, while kefiran is most approved due to its biodegradable and water-soluble nature (Piermaria et al., 2009). It is beneficial for human health as it has antimicrobial action and can therefore be used as a healing agent (Leite et al., 2013). Besides, its use is mostly studied in the food industry as a gelling agent, emulsifier and stabilizer (Piermaria et al., 2008). These films are capturing a great deal of attention because of their novel attributes such as biodegradability, stabilizing, biocompatibility, safety, emulsifying effects and water vapor permeability (Junior et al., 2020). Also they possess exceptional visual aspects and can be easily synthesized from glycerol (Hassan et al., 2019). Thus kefarin-based films are suitable for packaging food materials.

## 1.2.2 Polymers Sourced from Wood

Ligno-cellulosic fibers are 40–50% cellulose and 25–30% hemicelluloses. Cellulose-based films retain tensile strength, fine transparency, toughness and high surface gloss (Guzman-Puyol et al., 2019). Hemicelluloses are nebulous and complex heterogeneous polysaccharides having low thermal resistance (Bilal and Iqbal, 2020). These films are frangible, thus adding plasticizer enhances their elasticity, rigidity and allows minimum oxygen permeability (Zhang et al., 2020). Both hardwood (glucuronoxylan) and softwood hemicellulose films have an appealing role in packaging. However, these films are prone to aqueous uptake and more polymer content such as alginate enhances resistance against moisture uptake, but also it increases mechanical strength (Silva et al., 2015). Starch is the conventional plant-based polysaccharide used to produce bio-plastic films due to its cost efficiency, abundance and remarkable film-forming features (Hassan et al., 2018). Various starches either alone or in blended form are used as packaging material in order to increase the shelf life of a product (Thakur et al., 2019. Amalgamated coating of gelatin or corn starch plasticized with sorbitol, enhances post harvest storage of red crimson grapes. Starch molded with gelatin has better mechanical and solubility properties. It also has lower water vapor permeability in comparison to corn starch or modified blends (Fakhouri et al., 2015). Mango kernel starch is employed as a packaging envelope to extend the durability of tomatoes. This material proficiently slows down the ripening of tomatoes and it was examined by studying various chemical and physical parameters such as weight loss, hardiness, soluble solid concentration and tartaric acid content in fruit (Nawab et al., 2017).

## 1.2.3 Polymers Sourced from Proteins

Collagen and gelatin both are procured from animal origin. Collagen-based bio-films are produced by extrusion method whereas gelatin-based film synthesis demands wet process. Collagen-based bio-films have superior mechanical properties and tensile strength (Fadini et al., 2013). However, gelatin-based bio-film has poor barrier and mechanical characteristics, which display its hydrophilic nature (Ciannamea et al., 2018). Also, extrusion, compression and molding produce plasticized wheat gluten bio-films (Zubeldia et al., 2015). Disulfide, hydrophobic and hydrogen interactions are involved in synthesizing these bio-films. Transparency of film depends on gluten purity and casting medium employed during the synthesis process (Chiou et al., 2020). These bio-films manifest homogeneity, good gas barrier properties and mechanical strength (Mojumdar et al., 2011). Soy protein is used for the production of bio-films for packaging purposes. Bio-films generated from soy proteins are much clearer, smoother, elastic and economical compared to other protein based bio-films (Otoni et al., 2016). Bio-film synthesis from other proteins such as pistachio globulin protein, pea protein, pumpkin oil cake and canola protein has also been delineated in various studies (Zhang et al., 2018; Acquah et al., 2020; Umaraw and Verma, 2017; Popovic et al., 2012). Whey protein possesses extraordinary features and film-forming properties. These films are known for their unparalleled transparency, elasticity and oil or gas barrier characteristics at low humidity. However, bio-films produced from whey proteins demonstrate poor water barrier features. Blending oils or lipids during the synthesis process enhances water barrier properties (Bahram et al., 2014). Whey protein-based edible bio-films plasticized with glycerol were developed, which possess superior anti-microbial properties with extraordinary oxygen and water vapor permeability (Çakmak et al., 2020).

## 1.3 NANOTECHNOLOGY AND ADVANCEMENT OF BARRIER ATTRIBUTES

Excellent barrier properties serve as the requisite material required in food packaging. In contrast to ceramics, glass and metal, the polymers have poor barrier properties to liquids and gases and frequent improvement is always needed. Nanotechnology has emanated as an efficient method with a good success rate. Up-gradation of barrier properties is a pivotal topic and has been discussed in this chapter.

Nano-particles provide improvements in physical, barrier and mechanical characteristics and also serve as encapsulating agents for active portions. The inclusion of nano-materials in edible items acts as a delivery system for active components. This has resulted in the development of diverse nano-particle fortified edible materials also known as nano-composites (Chaturvedi and Dave, 2020; Shafiq et al., 2020; Liu et al., 2019).

### 1.3.1 Bio-Nanocomposites and Biodegradable Packaging

According to the market size and colossal amount of food materials consumed, the packaging industry has to take action more efficiently. Extensive research has been carried out in respect of eco-friendly packaging materials in order to reduce the effect on the environment caused by orthodox materials. And scientists came up with a solution that is bio-plastics. These are biodegradable polymers and are produced from renewable resources. Bio-plastics derived from agro wastes can be employed for food packaging. Polyethylene among the biodegradable polymers has hydro-degradation and oxo-degradation potential. Cellulose present in plants, bacteria and also in agro wastes is regarded as polymer. Using cellulose and its derivatives for food packaging gives us an insight of cellulose extraction from agro wastes, chemical alteration of cellulose within ionic medium, cellulose blending with other polymers, graft polymerization of cellulose (Zhao, 2018). Packaging must protect food items from physical damage and other spoilages such as dust, dirt and insects. The use of nano-fibers in packaging can bestow supportive physical properties. Also nano-materials viz, carbon nano-tubes and cellulose nano-fibers are stronger and stiffer compared to conventional materials (Siro and Plackett, 2010). Incorporating nano-materials into the conventional fillers may prove sufficient to improve the physical performance of the nano-composites (Smolander and Chaudhry, 2010). For extended shelf life and minimizing food waste, materials with high barrier properties are in huge demand. Exploiting nano-thin materials or nano-composites can render better barrier action.

Vacuum-based aluminium films on plastic are common barrier materials used for packaging food items such as coffee, snacks and confectionery. The aluminium films must be around 50 nm in thickness to qualify as nano-material. Lately, integration of nano-platelets from clays and polymers provides a complex path, which obstructs the passage of aroma, water and oxygen, thereby decreasing the rate of diffusion (Smolander and Chaudhry, 2010).

## 1.4 NANO-ENCAPSULATION TECHNIQUES

Nano-encapsulation is practiced for encasing solids, liquids and gaseous matter in minuscule capsules, which may release their content at a balanced rate when subjected to specific conditions of food items (Taylor et al., 2005). With the advent of nano-technology, the incessant research, emphasized on nano-encapsulation to achieve more durability, versatility and efficacy in food packaging. The aim here is to endow a critical perspective into nano-encapsulation pertaining to packaging in food businesses.

Spray drying is a broadly utilized micro-encapsulation approach in the food packaging industry, but, due to its low solubility and excessive water evaporation has led to it being a tiresome technique. Spray drying uses manifold coating materials including, soy/whey protein, sodium caseinate, gelatin, hydroxyl propyl methyl cellulose, carboxy methyl cellulose, fatty acids, chitosan, cholesterol, wax, etc (Heurtault et al., 2003; Kolanowski et al., 2004; Loksuwan, 2007). Nano-particle material can be stabilized by eliminating water into dissolvable dehydrated solid particles under the influence of various water soluble adjuvants such as drying auxiliaries (Tewa-Tagne et al., 2007). On the contrary, neither spray drying nor spray cooling exert water evaporation. Coacervation nano-encapsulation involves capsaicin capsules, which were formed by complex coacervation of tannins, gelatin and acacia and indicate possible anti-microbial application in the food industry (Xing et al., 2006). Self assembly of hydrolyzed milk protein viz., α-lactalbumin, results in straight and stabilized nano-tubes possessing the extraordinary feature of cavity. The distinctiveness of α-lactalbumin nano-tube renders it a potential encapsulating agent, such as, 8-nm cavity (Raviv et al., 2005). They also have the ability to resist conditions like pasteurization.

## 1.5 VOLATILE COMPOUNDS UTILIZED IN FOOD PACKAGING

Recently, a wide range of volatile compounds is reported as being used in food packaging industries. The infusion of these compounds into food packaging is increasing due to customer demand and advancement related to food safety and legislation. These chemical substances comprise aldehydes, terpenes and sulfur compounds. Monoterpenes are most commonly used as volatile compounds in food packaging and they are derived from essential oils of flowers, leaves, roots, fruits, seeds, bark, buds, herbs or woods (Kawacka et al., 2021). γ-terpinene has an application in food packaging and was incorporated into poly lactic acid and polybutylene succinate film to protect food from spoilage (Llana-Ruiz-Cabello et al., 2016). Eugenol belongs to the class of phenylpropanoids and possesses antioxidant and antibacterial action, it also shows hydrophobicity of packaging, thereby is suitable for food packaging (Navikaite et al., 2018). Although, use of eugenol for food packaging is limited due to its sensitivity towards heat, oxygen and light (Cheng et al., 2019) it is utilized as an active compound in food packaging because of its antifungal (Li et al., 2020), antioxidant (Melendez-Rodriguez et al., 2019) and antibacterial properties (Talon et al., 2019). Diallyl disulfide and allyl methyl sulfide are called metabolic products of sulfur comprising foods, chiefly garlic (Abe et al., 2020). The organosulfur compounds have high heat stability and powerful pungent odors (Nguyen and Yoshii, 2017. Garlic essential oils can possibly be encapsulated in packaging materials as microcapsules to retain freshness of food and it also shows antibacterial properties in the vapor phase (Zhang and Wang, 2019). Vanillin has a distinct odor and is obtained from vanilla beans, possessing potent antimicrobial action when infused into the coating or films of packaging materials (Sangsuwan and Sutthasupa, 2019; Buslovich et al., 2017). It is most extensively used as a food preservative or flavoring agent (Tajkarimi et al., 2010). Aldehydes are proven to enhance the shelf life of berries and fruits (Misran et al., 2015). Therefore, these compounds can be utilized by food industries to inhibit microbial growth and increase the shelf life of products.

## 1.6 FORTIFICATION OF BIO-PLASTIC FILMS

### 1.6.1 Bioactive Supplementation with Antioxidant Potential

Nowadays, people demand healthier and risk-free food, something which has urged scientists to come-up with new preservation strategies. Various approaches have been implemented to alleviate lipid peroxidation, for example, direct application of antioxidant to foods or their fusion with packaging materials (Kusznierewicz et al., 2020). Food items, notably fresh red meat or seafood cannot be packed without oxygen. Direct application of antioxidants to foods has drawbacks, as active ingredients deplete chemically and food starts to degrade almost mmediately(Navikaite-Snipaitiene et al., 2018). Currently, antioxidant packaging materials are produced in order to promote the stability of oxidation susceptible food items. To achieve this goal, antioxidants sourced from plants viz., essential oil are extensively studied. These essential oils possess remarkable antioxidant properties and antimicrobial properties (Deng et al., 2020). They are envisaged to help with less water uptake, due to their lipidic nature. They also facilitate the improvement of mechanical attributes of polymeric films such as tensile strength and optical structure (Iamareerat et al., 2018). Foline Ciocalteau assay is used to determine the phenolic content of biofilms. To measure the phenolic antioxidant activity synthetic radical, namely, 2,2-diphenyl-1-picrylhydrazyl, is used. Green tea decoction is blended with chitosan films, which demonstrate good antioxidant potential (Siripatrawan and Noipha, 2012). Reduction of $Fe^{+3}$ to $Fe^{+2}$ is precisely determined by ferric reducing antioxidant power (FRAP) at acidic pH. Chitosan bioplastic blended with thyme oil shows significant antioxidative strength, which was ascertained by this technique (Ruiz-Navajas et al., 2013). FRAP assay was employed to check the antioxidative capacity of gelatin biofilms fused with rosemary and oregano oils, and results revealed that oregano oil fusion based films are much more active than rosemary oil fusion based films (Gomez-Estaca et al., 2009). Soy protein based films exhibit heightened antioxidant potential when blended with red grapes concentrate (Ciannamea et al., 2016).

### 1.6.2 Bio-based Supplementation with Antimicrobial Potential

Anti-microbial agents are added in the packaging material to create an anti-microbial packaging, which encumbers microbial growth in packed food items. The packaging gives extended shelf life, enhanced quality and food safety by obviating pathogenic growth (Lavoine et al., 2014). Usage of anti-microbial agents for food packaging has reduced food borne epidemics (Kim and Rhee, 2016). And also, the consumer's preferences for the least processed and preservative free food made anti-microbial packaging far more enticing. The anti-microbial agents such as chitosan, essential oils and enzymes exploited for packaging are extracted from microbes, animals and plants (Qamar et al., 2020). Essential oils obtained from various plants are a superabundant source of phenolics and terpenoids, which are regarded as anti-microbial agents (Ruiz-Navajas et al., 2013). These essential oils act on microorganisms by diverse mechanisms such as destructing phospholipid bilayer of cell membrane, distorting enzyme structures and damaging genetic construction. Plant-based bioactive components have remarkable action against fungal and bacterial species (Fardioui et al., 2018). By various methods including disk diffusion, agar dilution and broth dilution method the anti-bacterial activity was affirmed. The anti-microbial action of garlic oil was determined by blending oil with bacterial inoculums (*Escherichia coli*, Staphylococcus aureus and *Salmonella typhimurium*). It was observed that garlic oil arrests bacterial growth and can be easily amalgamated into edible food packaging (Pranoto et al., 2005). Disk diffusion method is most suitable to screen anti-bacterial action of bio-films. Lately, garlic, rosemary and oregano oils were integrated into proteineous films and it was discerned that oregano oil based films have prominent anti-microbial action, whereas rosemary oil showed lesser anti-microbial action (Aldana et al., 2015).

### 1.7 CONCLUSION

The prospect of food packaging relies upon pioneering packaging that offers better protection, safety and stability. This chapter covers the novel and innovative strategies used in food packaging and discusses the major concerns about natural sources and their application in food packaging. Also nanotechnology based food packaging is elaborated on. The implementation of bio-plastic material in food packaging gives a novel, bio-degradable and eco-friendly substitute to petroleum and chemical based films. At the same time, these bio-based films are further being improved by incorporation of different bio-polymers in the form of blends and composites. It is explicit that bio-based packaging films offer multifaceted potential in the food packaging industry, however, certain improvements and storage tests are required in order to certify their use on a commercial scale. Also, critical efforts are being made to develop the right combination of materials, because the effectiveness of bio-polymeric coatings pivots upon the materials and polymers used to improve functional properties. Owing to nanotechnology use in active food packaging, the possibilities of improving food quality, efficiency and safety as a novel packaging material is just touching the surface. The outcome of nano-based packaging materials will need a paltry amount of material to prove effective. The technological advancements for nano-based food packaging achieved by developed countries must be adapted in developing nations too. Nano-encapsulation already being used in pharmaceuticals can also be adopted in the food industry by replacing conventional coatings for encapsulation of food contents.

## REFERENCES

Abe K, Hori Y, and Myoda T. 2020. Characterization of key aroma compounds in aged garlic extract. *Food Chem* 312: 126081.

Acquah C, Zhang Y, Dubé MA, and Udenigwe CC. 2020. Formation and characterization of protein-based films from yellow pea (Pisum sativum) protein isolate and concentrate for edible applications. *Curr Res Food Sci* 2: 61–69.

Ahvenainen R. ed. 2003. Novel food packaging techniques. *Elsevier.*

Aldana DS, Andrade-Ochoa S, Aguilar CN, Contreras-Esquivel JC, and Nevárez-Moorillón GV. 2015. Antibacterial activity of pectic-based edible films incorporated with Mexican lime essential oil. *Food Control* 50: 907–912.

Alshehrei F. 2017. Biodegradation of synthetic and natural plastic by microorganisms. *J Appl Environ Microbiol* 5: 8–19.

Ananda AP, Manukumar HM, Umesha S, Soumya G, Priyanka D, Kumar AM, Krishnamurthy NB, and Savitha KR. 2017. A relook at food packaging for cost effective by incorporation of novel technologies. *J Packag Technol Res* 1: 67–85.

Asgher M, Urooj Y, Qamar SA, and Khalid N. 2020. Improved exopolysaccharide production from Bacillus licheniformis MS3: optimization and structural/functional characterization. *Int J Biol Macromol* 151: 984–992.

Bahram S, Rezaei M, Soltani M, Kamali A, Ojagh SM, and Abdollahi M. 2014. Whey protein concentrate edible film activated with cinnamon essential oil. *J Food Process Preserv*, 38(3): 1251–1258.

Bilal M, and Iqbal HM. 2020. Ligninolytic enzymes mediated ligninolysis: an untapped biocatalytic potential to deconstruct lignocellulosic molecules in a sustainable manner. *Catal Lett* 150: 524–543.

Buslovich A, Horev B, Rodov V, Gedanken A, and Poverenov E. 2017. One-step surface grafting of organic nanoparticles: in situ deposition of antimicrobial agents vanillin and chitosan on polyethylene packaging films. *J Mater Chem B* 5: 2655–2661.

Çakmak H, Özselek Y, Turan OY, Fıratlıgil E, and Karbancioğlu-Güler F. 2020. Whey protein isolate edible films incorporated with essential oils: Antimicrobial activity and barrier properties. *Polym Degrad Stab* 179: 109285.

Cerqueira MA, Vicente AA, and Pastrana LM. 2018. Nanotechnology in food packaging: opportunities and challenges. In Cerqueira MAPR., Lagaron JM, Castro, L. M. P., & de Oliveira Soares, A. A. M. (Eds.). In Nanomaterials for Food Packaging: Materials, Processing Technologies, and Safety Issues (1–11). Elsevier Inc.

Chaturvedi S, and Dave PN. 2020. Application of nanotechnology in foods and beverages. In *Nanoengineering in the beverage industry* (137–162). Academic Press.

Chen GQ. 2010. Introduction of Bacterial Plastics PHA, PLA, PBS, PE, PTT, and PPP. In *Plastics from bacteria* (1–16). Springer.

Cheng J, Wang H, Kang S, Xia L, Jiang S, Chen M, and Jiang S. 2019. An active packaging film based on yam starch with eugenol and its application for pork preservation. *Food Hydrocoll* 96: 546–554.

Chiou BS, Cao T, Bilbao- Sainz C, Vega-Galvez A, Glenn G, and Orts W. 2020. Properties of gluten foams containing different additives. *Ind Crops Prod* 152: 112511.

Ciannamea EM, Castillo LA, Barbosa SE, and De Angelis MG. 2018. Barrier properties and mechanical strength of bio-renewable, heat-sealable films based on gelatin, glycerol and soybean oil for sustainable food packaging. *React Funct Polym* 125: 29–36.

Ciannamea EM, Stefani PM, and Ruseckaite RA. 2016. Properties and antioxidant activity of soy protein concentrate films incorporated with red grape extract processed by casting and compression molding. *LWT* 74: 353–362.

Darni Y, Dewi FY, and Lismeri L. 2017. Modification of Sorghum starch-cellulose bioplastic with Sorghum stalks filler. *Jurnal Rekayasa Kim Lingkung* 12: 22–30.

De Kort GW, Bouvrie LH, Rastogi S, and Wilsens CH. 2019. Thermoplastic PLA-LCP Composites: A Route toward Sustainable, Reprocessable, and Recyclable Reinforced Materials. *ACS Sustain Chem Eng* 8: 624–631.

Deng W, Liu K, Cao S, Sun J, Zhong B, and Chun J. 2020. Chemical composition, antimicrobial, antioxidant, and antiproliferative properties of grapefruit essential oil prepared by molecular distillation. *Molecules* 25: 217.

Eleftheriadou M, Pyrgiotakis G, and Demokritou P. 2017. Nanotechnology to the rescue: using nano-enabled approaches in microbiological food safety and quality. *Curr Opin Biotechnol* 44: 87–93.

Fadini AL, Rocha FS, Alvim ID, Sadahira MS, Queiroz MB, Alves RMV, and Silva LB. 2013. Mechanical properties and water vapour permeability of hydrolysed collagen–cocoa butter edible films plasticised with sucrose. *Food Hydrocoll* 30: 625–631.

Fakhouri FM, Martelli SM, Caon T, Velasco JI, and Mei LHI. 2015. Edible films and coatings based on starch/gelatin: Film properties and effect of coatings on quality of refrigerated Red Crimson grapes. *Postharvest Biol Technol* 109: 57–64.

Fardioui M, Kadmiri IM, and Bouhfid R. 2018. Bio-active nanocomposite films based on nanocrystalline cellulose reinforced styrylquinoxalin-grafted-chitosan: Antibacterial and mechanical properties. *Int J Biol Macromol* 114: 733–740.

Gómez-Estaca J, Bravo L, Gómez-Guillén MC, Alemán A, and Montero P. 2009. Antioxidant properties of tuna-skin and bovine-hide gelatin films induced by the addition of oregano and rosemary extracts. *Food Chem* 112: 18–25.

Guilbert S, Gontard N, and Cuq B. 1995. Technology and applications of edible protective films. *Packag Technol Sci* 8: 339–346.

Guzman-Puyol S, Ceseracciu L, Tedeschi G, Marras S, Scarpellini A, Benítez JJ, Athanassiou A, and Heredia-Guerrero JA. 2019. Transparent and robust all-cellulose nanocomposite packaging materials prepared in a mixture of trifluoroacetic acid and trifluoroacetic anhydride. *J Nanomater* 9: 368.

Hassan AA, Abbas A, Rasheed T, Bilal M, Iqbal HM, and Wang S. 2019. Development, influencing parameters and interactions of bioplasticizers: An environmentally friendlier alternative to petro industry-based sources. *Sci Total Environ* 682: 394–404.

Hassan B, Chatha SAS, Hussain AI, Zia KM, and Akhtar N. 2018. Recent advances on polysaccharides, lipids and protein based edible films and coatings: A review. *Int J Biol Macromol* 109: 1095–1107.

Heurtault B, Saulnier P, Pech B, Proust J, and Benoit JP. 2003. Physico-chemical stability of colloidal lipid particles. *Biomaterials* 24: 4283–4300.

Huang HJ, and Ramaswamy S. 2013. Overview of biomass conversion processes and separation and purification technologies in biorefineries. *Sep Purif Technol biorefineries* 1–36.

Iamareerat B, Singh M, Sadiq MB, and Anal AK. 2018. Reinforced cassava starch based edible film incorporated with essential oil and sodium bentonite nanoclay as food packaging material. *J Food Sci Technol* 55: 1953–1959.

Iqbal HM, Rasheed T, and Bilal M. 2018. Design and processing aspects of polymer and composite materials. *Green and Sustainable Advanced Materials: Processing and Characterization* 1: 155–189.

Jain R, and Tiwari A. 2015. Biosynthesis of planet friendly bioplastics using renewable carbon source. *J Environ Health Sci Eng* 13: 1–5.

Jiang T, Duan Q, Zhu J, Liu H, and Yu L. 2020. Starch-based biodegradable materials: Challenges and opportunities. *Adv Ind Eng Polym Res* 3: 8–18.

Júnior LM, Vieira RP, and Anjos CAR. 2020. Kefiran-based films: Fundamental concepts, formulation strategies and properties. *Carbohydr Polym* 246: 116609.

Kawacka I, Olejnik-Schmidt A, Schmidt M, and Sip A. 2021. Natural plant-derived chemical compounds as Listeria monocytogenes inhibitors in vitro and in food model systems. *Pathogens* 10: 12.

Kawaguchi H, Hasunuma T, Ogino C, and Kondo A. 2016. Bioprocessing of bio-based chemicals produced from lignocellulosic feedstocks. *Curr Opin Biotechnol* 42: 30–39.

Kim SA, and Rhee MS. 2016. Highly enhanced bactericidal effects of medium chain fatty acids (caprylic, capric, and lauric acid) combined with edible plant essential oils (carvacrol, eugenol, β-resorcylic acid, trans-cinnamaldehyde, thymol, and vanillin) against Escherichia coli O157: H7. *Food Control* 60: 447–454.

Kolanowski W, Laufenberg G, and Kunz B. 2004. Fish oil stabilisation by microencapsulation with modified cellulose. *Int J Food Sci Nutr* 55: 333–343.

Kusznierewicz B, Staroszczyk H, Malinowska-Pańczyk E, Parchem K, and Bartoszek A. 2020. Novel ABTS-dot-blot method for the assessment of antioxidant properties of food packaging. *Food Packag Shelf Life* 24: 100478.

Lavoine N, Givord C, Tabary N, Desloges I, Martel B, and Bras J. 2014. Elaboration of a new antibacterial bio-nano-material for food-packaging by synergistic action of cyclodextrin and microfibrillated cellulose. *Innov Food Science Emerg Technol* 26: 330–340.

Leite AMDO, Miguel MAL, Peixoto RS, Rosado AS, Silva JT, and Paschoalin VMF. 2013. Microbiological, technological and therapeutic properties of kefir: a natural probiotic beverage. *Braz J Microbiol* 44: 341–349.

Li Y, Dong Q, Chen J, and Li L. 2020. Effects of coaxial electrospun eugenol loaded core-sheath PVP/shellac fibrous films on postharvest quality and shelf life of strawberries. *Postharvest Biol Technol* 159: 111028.

Liu X, Zhang B, Sohal IS, Bello D, and Chen H. 2019. Is "nano safe to eat or not"? A review of the state-of-the art in soft engineered nanoparticle (sENP) formulation and delivery in foods. *Adv Food Nutr Res* 88: 299–335.

Llana-Ruiz-Cabello M, Pichardo S, Bermúdez JM, Baños A, Núñez C, Guillamón E, Aucejo S, and Cameán AM. 2016. Development of PLA films containing oregano essential oil (Origanum vulgare L. virens) intended for use in food packaging. *Food Addit Contam: Part A* 33: 1374–1386.

Loksuwan J. 2007. Characteristics of microencapsulated β-carotene formed by spray drying with modified tapioca starch, native tapioca starch and maltodextrin. *Food hydrocoll* 21: 928–935.

Luzi F, Torre L, Kenny JM, and Puglia D. 2019. Bio-and fossil-based polymeric blends and nanocomposites for packaging: Structure–property relationship. *Materials* 12: 471.

Melendez-Rodriguez B, Figueroa-Lopez KJ, Bernardos A, Martínez-Máñez R, Cabedo L, Torres- Giner S, and M Lagaron J. 2019. Electrospun antimicrobial films of poly (3-hydroxybutyrate-co-3-hydroxyvalerate) containing eugenol essential oil encapsulated in mesoporous silica nanoparticles. *Nanomaterials* 9: 227.

Misran A, Padmanabhan P, Sullivan JA, Khanizadeh S, and Paliyath G. 2015. Composition of phenolics and volatiles in strawberry cultivars and influence of preharvest hexanal treatment on their profiles. *Can J Plant Sci* 95: 115–126.

Mojumdar SC, Moresoli C, Simon LC, and Legge RL. 2011. Edible wheat gluten (WG) protein films: preparation, thermal, mechanical and spectral properties. *J Therm Anal Calorim* 104: 929–936.

Nampoothiri KM, Nair NR, and John RP. 2010. An overview of the recent developments in polylactide (PLA) research. *Bioresour Technol* 101: 8493–8501.

Navikaite- Snipaitiene V, Ivanauskas L, Jakstas V, Rüegg N, Rutkaite R, Wolfram E, and Yildirim S. 2018. Development of antioxidant food packaging materials containing eugenol for extending display life of fresh beef. *Meat Sci* 145: 9–15.

Nawab A, Alam F, and Hasnain A. 2017. Mango kernel starch as a novel edible coating for enhancing shelf-life of tomato (Solanum lycopersicum) fruit. *Int J Biol Macromol* 103: 581–586.

Nguyen TVA, and Yoshii H. 2017. Encapsulation of Allyl Sulfide with Middle–Chain Triglyceride Oil and Cyclodextrin by Spray Drying. *Japan J Food Eng* 16480.

Otoni CG, Avena- Bustillos RJ, Olsen CW, Bilbao-Sáinz C, and McHugh TH. 2016. Mechanical and water barrier properties of isolated soy protein composite edible films as affected by carvacrol and cinnamaldehyde micro and nanoemulsions. *Food Hydrocoll* 57: 72–79.

Pal AK, and Katiyar V. 2017. Thermal degradation behaviour of nanoamphiphilic chitosan dispersed poly (lactic acid) bionanocomposite films. *Int J Biol Macromol* 95: 1267–1279.

Paletta A, Leal Filho W, Balogun AL, Foschi E, and Bonoli A. 2019. Barriers and challenges to plastics valorisation in the context of a circular economy: Case studies from Italy. *J Clean Prod* 241: 118149.

Pan K, and Zhong Q. 2016. Organic nanoparticles in foods: fabrication, characterization, and utilization. *Annu Rev Food Sci Technol* 7: 245–266.

Park JH, Koo MS, Cho SH, and Lyu MY. 2017. Comparison of thermal and optical properties and flowability of fossil-based and bio-based polycarbonate. *Macromol Res* 25: 1135–1144.

Piermaria JA, Mariano L, and Abraham AG. 2008. Gelling properties of kefiran, a food-grade polysaccharide obtained from kefir grain. *Food hydrocoll* 22: 1520–1527.

Piermaria JA, Pinotti A, Garcia MA, and Abraham AG. 2009. Films based on kefiran, an exopolysaccharide obtained from kefir grain: Development and characterization. *Food hydrocoll* 23: 684–690.

Popović S, Peričin D, Vaštag Ž, Lazić V, and Popović L. 2012. Pumpkin oil cake protein isolate films as potential gas barrier coating. *J Food Eng* 110: 374–379.

Pranoto Y, Salokhe VM, and Rakshit SK. 2005. Physical and antibacte rial properties of alginate-based edible film incorporated with garlic oil. *Food Res Int* 38: 267–272.

Qamar SA, Asgher M, and Bilal M. 2020. Immobilization of alkaline protease from Bacillus brevis using Ca-alginate entrapment strategy for improved catalytic stability, silver recovery, and dehairing potentialities. *Catal Lett* 150: 3572–3583.

Raviv U, Needleman DJ, Li Y, Miller HP, Wilson L, and Safinya CR. 2005. Cationic liposome–microtubule complexes: Pathways to the formation of two-state lipid–protein nanotubes with open or closed ends. *Proc Natl Acad Sci* 102: 11167–11172.

Ruiz-Navajas Y, Viuda-Martos M, Sendra E, Perez-Alvarez JA. Fernández-López J. 2013. In vitro antibacterial and antioxidant properties of chitosan edible films incorporated with Thymus moroderi or Thymus piperella essential oils. *Food Control* 30: 386–392.

Sangsuwan J, and Sutthasupa S. 2019. Effect of chitosan and alginate beads incorporated with lavender, clove essential oils, and vanillin against Botrytis cinerea and their application in fresh table grapes packaging system. *Packag Technol Sci* 32: 595–605.

Shafiq M, Anjum S, Hano C, Anjum I, and Abbasi BH. 2020. An overview of the applications of nanomaterials and nanodevices in the food industry. *Foods* 9: 148.

Shit SC, and Shah PM. 2014. Edible polymers: challenges and opportunities. *J Polym 2014*.

Silva TCF, Silva Deusanilde, and Lucia LA. 2015. The Multifunctional Chemical Tunability of Wood-Based Polymers for Advanced Biomaterials Applications. In *Green Biorenewable Biocomposites: From Knowledge to Industrial Applications* (427–459).

Siripatrawan U, and Noipha S. 2012. Active film from chitosan incorporating green tea extract for shelf life extension of pork sausages. *Food hydrocoll* 27: 102–108.

Siró I, and Plackett D. 2010. Microfibrillated cellulose and new nanocomposite materials: a review. *Cellulose* 17: 459–494.

Smolander M, and Chaudhry Q. 2010. Nanotechnologies in food packaging. *Nanotechnologies in food* 14: 86–101.

Tajkarimi MM, Ibrahim SA, and Cliver DO. 2010. Antimicrobial herb and spice compounds in food. *Food Control* 21: 1199–1218.

Talón E, Vargas M, Chiralt A, and González-Martínez C. 2019. Eugenol incorporation into thermoprocessed starch films using different encapsulating materials. *Food Packag Shelf Life* 21: 100326.

Taylor TM, Weiss J, Davidson PM, and Bruce BD. 2005. Liposomal nanocapsules in food science and agriculture. *Crit Rev Food Sci Nutr* 45: 587–605.

Tewa-Tagne P, Briançon S, and Fessi H. 2007. Preparation of redispersible dry nanocapsules by means of spray-drying: development and characterisation. *Eur J Pharm Sci* 30: 124–135.

Thakur R, Pristijono P, Scarlett CJ, Bowyer M, Singh SP, and Vuong QV. 2019. Starch-based films: Major factors affecting their properties. *Inl J Biol Macromol* 132: 1079–1089.

Tsagkaris AS, Tzegkas SG, and Danezis GP. 2018. Nanomaterials in food packaging: state of the art and analysis. *J Food Sci Technol* 55: 2862–2870.

Umaraw P, and Verma AK. 2017. Comprehensive review on application of edible film on meat and meat products: An eco-friendly approach. *Crit Rev Food Sci Nutr* 57: 1270–1279.

Xing F, Cheng G, and Yi K. 2006. Study on the antimicrobial activities of the capsaicin microcapsules. *J Appl Polym Sci* 102: 1318–1321.

Zhang H, and Wang J. 2019. Constituents of the essential oils of garlic and citronella and their vapor-phase inhibition mechanism against S. aureus. *Food Sci Technol Res* 25: 65–74.

Zhang X, Xiao N, Chen M, Wei Y, and Liu C. 2020. Functional packaging films originating from hemicelluloses laurate by direct transesterification in ionic liquid. *Carbohyd polym* 229: 115336.

Zhang Y, Liu Q, and Rempel C. 2018. Processing and characteristics of canola protein-based biodegradable packaging: A review. *Crit Rev Food Sci Nutr* 58: 475–485.

Zhao CT. 2018. Food packaging. *Johnson Matthey Technol Rev* 62: 74–80.

Zubeldía F, Ansorena MR, and Marcovich NE. 2015. Wheat gluten films obtained by compression molding. *Polym Test* 43: 68–77.

CHAPTER 2

# Contaminants and Their Control Measures to Maintain Hygiene in Food Processing

**Varsha Singh*** **and Amit Wadehra**
Chitkara University, Punjab
*Corresponding author email: varshasingh213@gmail.com

## CONTENTS

| | | |
|---|---|---|
| 2.1 | Introduction | 18 |
| 2.2 | Food Processing Steps | 19 |
| | 2.2.1 Unprocessed Material | 19 |
| | 2.2.2 Contamination of Food during Transportation | 20 |
| | 2.2.3 Infectivity Due to Heat Treatment | 20 |
| | 2.2.4 Food Packaging and Storage | 21 |
| |     2.2.4.1 Food Packaging | 21 |
| |     2.2.4.2 Storage of Food | 22 |
| | 2.2.5 Food Safety Hazard | 22 |
| 2.3 | Methods for Ensuring Food Safety and Hygiene | 23 |
| | 2.3.1 Good Manufacturing Practices (GMP) | 23 |
| | 2.3.2 Sanitation Standard Operating Procedure (SSOP) | 23 |
| | 2.3.3 Good Hygiene Practices | 24 |
| | 2.3.4 Hazard Analysis and Critical Control Points (HACCP) | 24 |
| |     2.3.4.1 Stage 1: Assemble the HACCP Team | 25 |
| |     2.3.4.2 Stage 2: Describe Product/ Process | 26 |
| |     2.3.4.3 Stage 3: Identify Intended Use | 27 |
| |     2.3.4.4 Stage 4: Construct Process Flow Diagrams | 27 |
| |     2.3.4.5 Stage 5: Confirm Accuracy of Process Flow Diagrams | 27 |
| |     2.3.4.6 Stage 6: Conduct a Hazard Analysis | 27 |

DOI: 10.1201/9781003258568-2

2.3.4.7　Stage 7: Determine Critical Control Points ...................... 28
　　　2.3.4.8　Stage 8: Establish Critical Limits for Each CCP ............... 28
　　　2.3.4.9　Stage 9: Establish a Monitoring System
　　　　　　　 for Each CCP ................................................................. 28
　　　2.3.4.10　Stage 10: Establish Corrective Actions ............................ 28
　　　2.3.4.11　Stage 11: Establish Verification Procedures ..................... 29
　　　2.3.4.12　Stage 12: Establish Documentation and
　　　　　　　　Record-Keeping ............................................................ 29
2.4　Conclusion ............................................................................................ 29
References ....................................................................................................... 30

## 2.1 INTRODUCTION

Food hygiene refers to the circumstances and procedures that must be followed from the point of manufacturing to the point of utilization to ensure food safety. The safety of the food produced is vital for every food manufacturing process (Kamboj et al., 2020). When there are no harmful substances in food, it is considered safe and hygienic. Foods and liquids can get polluted, rendering them unfit for ingestion.

Food quality, safety laws and regulations must be strengthened to maintain the equality of buying and selling of food that assures customers' wellbeing since food production, shipping, and supply become gradually more universal. On the other hand, microbial and chemical pollutants frequently cause food safety issues (Jing et al., 2015). Among the many sources of hazardous compounds discovered in raw materials during food processing, packaging, and storage are toxicants, naturally occurring pollutants such as toxins, pesticides, and packing substrates. Toxicants are compounds that are produced by the breakdown of organic matter. Groh et al. (2017) assert that certain dietary pollutants, such as pesticide residues and migrants, may have a detrimental effect on the individual gastrointestinal microbiota. These pollutants require research and analytical methods to identify and detect them. Contaminants pose a problem during food processing, regardless of their source, whether they come from food products, the manufacturing process, or material used for packaging. Safe food is inextricably linked to public wellbeing; hence technical and realistic strategies to avoid and diminish food pollutants throughout manufacturing must be implemented. Despite the fact that many countries have made tremendous strides in improving food safety standards, the prevalence of infectious diseases remains high around the world. Foodborne illness prevention and minimization are becoming increasingly difficult as new risk factors emerge in the food supply. Foodborne infections have become more common as a result of economic globalization (Guerrini, 2005). Food logistics has a huge disparity in its food protection monitoring system. Unexpected food safety dangers and concerns have arisen due to altering distinctiveness in the food supply chain. As a result, statistics and information on potential pollutants at each stage of food processing are critical. Chemical

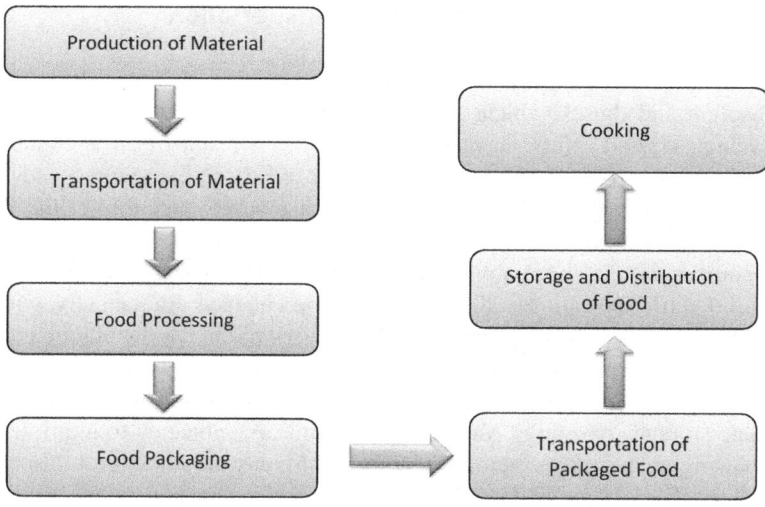

**Figure 2.1** Food processing stages.

pollutants in food processing are discussed in this overview and techniques to mitigate and prevent them (see Figure 2.1).

## 2.2 FOOD PROCESSING STEPS

### 2.2.1 Unprocessed Material

Transport, handling, storing, and preparing food all have an effect on food safety. However, it is also influenced by the main surroundings since the beginning points to the consumable end (Brambilla and Testa, 2014; Djekic et al., 2018). When food materials come into contact with the environment, hazardous environmental contaminants such as heavy metals and organic pollutants can build up. The crucial step and front line of defence for ensuring food safety is controlling the food's primary environmental quality from the point of origin.

Food contamination can be caused by industrial growth, advancements in agrochemicals, or urban activity. Insecticides and agrochemicals, which can be harmful to humans if consumed. Pesticide residues were observed in vegetables and fruits in several studies (Kobayashi et al., 2011) and a few imitative with potentially harmful effect, like organochlorine metabolites pesticide, which were discovered in fatty foods (Chung and Chen, 2011). As reported by the ATSDR (2011), toxic heavy metals such as cadmium, lead, mercury, and arsenic can be present in the environment's water, soil, and air (Zukowska and Biziuk, 2008) and they can be transferred to food through contact with these elements. Heavy metals have been detected in various foods, including honey, spinach, potatoes, seafood, and tea.

## 2.2.2 Contamination of Food during Transportation

Contamination of food can also occur when it is being transported. It can be caused by gasoline and diesel vehicle emissions and cross-contamination in food transport vehicles. Food safety may be jeopardized as a result of cross-contamination. Containers infected with fungicides that were used for the storage and transportation of packaging materials, were blamed for a severe sickness of the European Economic Community in 1999. Cross-contamination from disinfection chemicals or other sources has also been a problem on long-distance transport ships (Nerin et al., 2007). Food transportation and distribution have changed dramatically, with certain corporations serving as the primary mode of distribution. On the other hand, mainly fast deliverance firms do not hoard foods separately after receiving an order but rather mix them up with other items. As a result, food contamination occurs due to a lack of adequate storage conditions. Various novel items are subjected to multiple freezing and thawing cycles, and whereas there is no obvious modification in aspects, their quality has been harmed. Certain businesses are increasingly concentrating their efforts on the development of frozen meat storage and public cold storage, however, they frequently overlook the development of cold storage facilities for fresh fruit and vegetables. For the entire food supply chain, cold storage is a critical infrastructure. When businesses fail to build cold storage units and equipment, they cannot ensure that agricultural products are kept at the proper temperature during their journey.

## 2.2.3 Infectivity Due to Heat Treatment

When high heat cooking is combined with external influences, hazardous chemicals are formed, which can harm food quality and safety. Processed foods may be unsafe for consumption, especially if they are subjected to a hazard during food processing, such as during heating, braising, roasting, canning, or oxidation. Frying is perhaps the most dangerous cooking method because it has the potential to introduce an extensive range of contaminated compound into the food. Reactions of oxidized frying oil with proteins and other sulphur and nitrogen components yield flavor chemicals. Color and flavors are also liberated from the meal into the frying oil, resulting in off-flavors and off-coloration. Aside from that, pigments contained in frying oil can be absorbed into the surface of fatty food, which causes the meal to change color. Also, essential contaminants resulting from heating processes are acrylamide and its precursors, which are both toxic. Nitrosamines are a processing contaminant that can be formed due to the contact of natural food components with food additives while the food is being heated. Because of the drying and roasting processes that use direct heat, the presence of nitrosodimethylamine in certain foods has been discovered. The amount of nitrosamine created during vapor or boiling cooking (which indicates lower temperatures, such as 100 degrees Celsius) is substantially less than the amount formed during frying, roasting, or grilling. Polycyclic aromatic hydrocarbons (PAHs), which are found in foods such as grilled and smoked meats, ethyl carbonate, and other products, and furan derivatives, which are found in a variety of heat-treated foods, particularly coffee and canned and jarred foods, are among the other processing contaminants formed during heating. In addition to ascorbic acid, furan can be formed from a variety of predecessors, including carbohydrate breakdown and amino acid breakdown as well as fatty acid oxidation. Furan is a flavoring agent that can be produced by a variety of processes in the food industry. In the absence of fat, the production of mutagens is significantly reduced. Fats and iron salts were mixed with glycerol, glucose, and creatinine, and then heated, simulating the Maillard reaction, which leads to several flavor precursors. In the presence of oxidized fats and iron salts, the reaction was accelerated, but it was not inhibited by tocopherol.

Frying fats, in addition to decomposition products resulting from nitrogen-free lipid-hydroperoxide breakdown, contribute to some of the mutagenesis activity found in frying fats, even when no fat is used as a substrate. Cooking foods in the microwave is becoming a more popular method of heating foods in the home and in certain industrial settings. A typical property of microwave cooking is that the food is cooked in the packing material (wrapping film or container) in the microwave oven (Nerin et al., 2003) is a common characteristic of microwave cooking. Moldable packaging materials include plastics, paperboard, and composites, and many of the elements of these materials can migrate from the package into the food when microwave cooking is performed and those rates are accelerated. Because of this, the quality and safety of food is being jeopardized (Ehlert et al., 2008). Microwaves can also cause migration of migrants or polymers to degrade and create hot spots, which can cause immigration to reach an upper point, which is less predictable because of the mass high temperature (Nerin et al., 2002).

### 2.2.4 Food Packaging and Storage

#### 2.2.4.1 Food Packaging

Compounds are incorporated into the materials used to manufacture packaging in order to achieve better performance and storage. However, under certain conditions, these chemicals may dissolve and migrate into food. As a result, the migration of packaging materials can raise concerns about food safety. There are several key advantages associated with paper packaging materials, such as versatility, dependability, sustainability, and reusability. There is a possibility that paper brighteners and colorants could migrate from food processing and storage to food (Xue et al., 2016a; Xue et al., 2016b; Li et al., 2018). Metal packaging materials are an outstanding product, improving the shelf life of food products while also reducing the impact of water and sunlight on the products' appearance and flavor.

On the other hand, metal materials are susceptible to destruction by matter found in food, making it essential to ensure they remain rust free. Lead and chromium may transfer from metal packaging materials to food products during food storage (Huang et al., 2014; Shu-Juan et al., 2014). It is common practice to add antioxidants and plasticizers to plastic packaging materials, which can leach into food. Apart from carcinogenesis, the majority of these additives possess adverse health consequences such as impaired growth and reproduction. While adding phthalate esters to plastics may boost their elastic properties and thermal stability, PAEs are toxic chemicals that have been shown to injure an animal's fertility. Few aspects such as the packaging and food contact area, the amount of time spent in contact with food, and the storage period may all have an effect on the PAEs' migration into foods during storage and processing (Liu, 2016; Jiao et al., 2017).

### 2.2.4.2 Storage of Food

When food is stored, the conditions in which it is kept have an effect on both the quality and the safety of the food. Microorganisms inevitably affect food during storage, transportation, and distribution. The scientific community has paid close attention to mycotoxins (Patriarca and Fernandez Pinto, 2017). Agri products must be piled up in climate-controlled conditions. Pesticide residue concentrations in food are affected by a variety of factors, including storage dampness and temperature. For instance, after eight months of storage, wheat retains 80% of malathion. Many such bacterial enzymes degrade harvest residue on grains; the octanol-water concentration gradient of pesticides is temperature and humidity dependent (Balinova et al., 2006; Uygun et al., 2008; Uygun et al., 2009). Crops stored in huge amounts may cause pesticides to degrade more rapidly (Lalah and Wandiga, 2002).

The acids inherent in fruits cause pesticide residues to break down more quickly than those found in grains (Rasmusssen et al., 2003). During storage, transportation, and distribution, animal protein foods are typically frosty or chilled. Due to the high protein content and prooxidant part in meat, the protein endures a varying extent of oxidation throughout refrigeration, ensuing in a loss of protein content and quality. For instance, oxidative strain induces carbonylation in pullet protein, a process allied with muscle type and mellowness (Estevez, 2015). This is supported by Soyer et al. (2010), who state that animal species and fat content are key influencing factors on protein oxidation, storage temperature, and time. Throughout the manufacturing and marketing processes, meat products are repeatedly frozen, which accelerates the oxidation of protein carbonyl groups. Intake of oxidized compounds has been shown to increase oxidative identifiers in muscle and blood and the possibility of increased oxidation in animal tissues (Zhang et al., 2011). The oxidized protein produces specific metabolites in the presence of colonic bacteria, which may amplify the menace of malignant cells and colitis. As a result, safe food stocking, shipping, and dispersal must be practiced (Wei et al., 2017; Wu, 2017).

### 2.2.5 Food Safety Hazard

Contaminants that pose a risk to food safety are those that effectively make a food item unsuitable for manufacturing. Codex (1997) defines hazard as "Hazard: a biological, chemical or physical agent in, or condition of, food with the potential to cause an adverse health effect". Any food product may be contaminated during preparation or while being transported, especially if hazardous compounds are added into the mix. It is critical to comprehend the potential risk associated with the chosen ingredient types or the processing environment. This enables the project team to determine the most effective methods for mitigating such risks, whether by stopping their access into the practice, ruining, or decreasing infectivity to an acceptable point. This is no longer a potential food hazard due to the way it is processed. This data on potential hazards and suggested preventative measures should be integrated with the prerequisite clinical best practices program and HACCP system to make sure that daily checking is created in the production process.

## 2.3 METHODS FOR ENSURING FOOD SAFETY AND HYGIENE

Good manufacturing practices.
Sanitation standard operating procedures.
Good hygiene practices.
Hazard analysis & critical control points.

### 2.3.1 Good Manufacturing Practices (GMP)

Good Manufacturing Practices refers to the procedures used in food, drugs, and medical equipment. The four principles upon which GMPs are based are: excluding undesired and foreign substances, inhibiting microorganisms, and destroying microbial pathogens. GMP is comprised of the following components: the facility and its atmosphere, the employees, sanitizing and cleaning methods, kitchen utensils, systems and procedures, and storage and transport. The GMP program's analysis and control of these elements are aimed at ensuring the manufacturing premium food. GMPs are one method of preventing the spread of food borne illness. Manufacturing companies that have implemented GMP programmes have seen the following benefits: increased food quality; healthier goods; reduced customer complaints; a more pleasant, cleaner, and healthier workplace; increased employee performance and morale; and enhanced psychosocial factors (Da Cruz et al., 2006). GMP execution is a continuous practice that is guided by the Plan, Do, Check, and Act cycle. Completion of GMP can be alienated into four stages: initial prognosis, development of the blueprint; resolution of discrepancies; and reevaluation of remedial action. Initial prognosis and reevaluation of remedial action are typically conducted through facility inspections using a checklist developed under the country's GMP regulations. While road maps can be generated following an inspection, corrective action frequently requires the prioritization of areas based on available funds and labors within the organization (Dias et al., 2012). These priority areas can be determined through episodic inspection conducted by government bureau or domestic controls.

### 2.3.2 Sanitation Standard Operating Procedure (SSOP)

SSOPs are composing techniques designed and implemented in a unit to preclude goods from being contaminated directly or from being adulterated. According to Cruz et al. (2006), SSOPs contain a detailed description of the specific practices needed to keep equipment and utensils free of microbial pathogens and with a minimum of decaying microbiota, thereby extremely important to maintain foodstuffs that come into contact with this gear and these utensils. The unit must keep these composed techniques on file and make them available upon request to regulatory or government bureau. The following SSOP central directives have been established in light of possible contamination sources: from raw to cooked food contamination; the product in contact with contaminated water and perhaps other potentially hazardous substances; contact with substances other than food, such as pesticides; contact with aerosolized substances; disease or insufficient hygiene on the part of the handler; pest control and unfamiliar material. According to Cunha et al. (2015) kitchen workers who have not participated in food safety training have higher tension, anxiety and lesser understanding on issues related to food safety, which means that education can improve knowledge, while empowering food handlers to improve self effectiveness and reduce anxiety and stress.

### 2.3.3 Good Hygiene Practices

GHPs are methods and procedures that adhere to best practice standards in order to be considered effective (British Retail Consortium, 2011). According to the European Commission, food hygiene is the technique and conditions necessary to minimize dangers and concentrate on maintaining the fitness of foodstuffs for human consumption, taking into consideration the intended use of the foodstuff. When it comes to food safety and quality management systems, good hygiene practices are typically referred to as the foundational measures on which other systems are built. They encompass a comprehensive file of trial, including personal hygiene and training for employees. According to the Food Standards Agency (2009), food hygiene training is a mandatory requirement under the law to ensure that safe food preparation practices are followed and maintained. The main reasons why sanitation training failed were how it was carried out, the demographics of the trainee and their ability to learn, no one supervising after training, a lack of resources to reinforce information in low-income areas (Gilling et al., 2001). Feglo et al. (2004) emphasized the importance of training and surveillance in areas where the establishment and design of permissible infrastructure facilities could take years due to cost. Feglo and Sakyi (2012) emphasized the critical nature of hygiene exercise in the food business.

### 2.3.4 Hazard Analysis and Critical Control Points (HACCP)

This is a structured collection of tasks used to constantly monitor the production of food in order to improve food security and avoid food product changes. The system is predicated on applying control practices in specific production steps where health hazards are more likely to occur. Good manufacturing practices and sanitation standard operating procedures are precondition programs for implementing HACCP in food industries. This array of food industry segments covers everything from building structure and keeping food safe to water treatment systems, cleaning routines, sanitation equipment, and quality assurance standards for products and raw materials (Barendz, 1998). At every stage of the food chain, from raw material production to final product, the system is applied and includes aspects related to consumer demands, such as processed foods that are not harmful to their health (World Health Organization, 1997). According to Bryan (1992), the HACCP system is based on predefined concepts and terms.

- Hazard: Food contaminated in an unacceptably biological, physical, or chemical manner, rendering it unfit for consumption.
- Threat: The projected possibility of a hazard occurrence.
- Critical Control Point: Preventive measures are taken when a certain product is being made in order to ensure the product is kept under supervision and any potential health concerns are eliminated.
- Critical limit: The value or characteristic assigned to each variable associated with a critical point. Non-adherence exposes consumers to health risks. Limits are established based on criteria such as guidelines, particular literature, industry-specific knowledge, past surveys, company policies, and other resources.
- Emendation: Instant and relevant corrective action to be taken when critical limits are violated.
- Affirmation: Conducting additional tests or an assessment of HACCP records to verify the system's integrity.
- Resoluteness: Providing a list of sequences that have been applied to identify a specific danger in a substance, food item, or process step.

HACCP implementation should adhere to seven fundamental principles (WHO, 1997). These 12 phases were laid out in the Codex Alimentarius (1997) to set up a seven-step programme that incorporates these principles:

Stage 1:- Assemble the HACCP Team
Stage 2:- Describe Product
Stage 3:- Identify Intended Use
Stage 4:- Construct Flow Diagram
Stage 5:- On-Site Confirmation of Flow Diagram
Stage 6:- List All Potential Hazards, Conduct a Hazard Analysis & Consider Control Measures.
Stage 7:- Determine CCPs
Stage 8:- Establish Critical Limits of Each CCP
Stage 9:- Establish a Monitoring System for Each CCP
Stage 10:- Establish Corrective Actions
Stage 11:- Establish Verification Procedures
Stage 12:- Establish Documentation & Record Keeping

### 2.3.4.1 Stage 1: Assemble the HACCP Team

In order to help guarantee food safety hazards and controls are given adequate consideration across all relevant disciplines, HACCP employs multidisciplinary staff who bring relevant skills, knowledge, and experience together to enable collective knowledge of how to minimize food safety hazards to the public's health. The whole interdisciplinary nature of the HACCP team is regarded as one of HACCP's patent assets. The HACCP team's crucial expertise includes the following:

- Knowledge of on-site production processes, raw materials, and products.
- Equipment knowledge and experience, including how the equipment operates to achieve operating conditions and the most common failure mechanism.
- Awareness of potential hazards and suitable regulatory systems, such as product development safety standards and control systems, as well as the ability to affirm all required regulatory standards.
- Working understanding of the HACCP concepts and how to put them into practice is required.

The HACCP team leader and an administrator/scribe should be designated to assist with documenting food processing activities. To function effectively as a HACCP team, all team members must understand how HACCP principles are applied. A hands-on training program that incorporates both HACCP theory and application should be used to educate the entire team in order to obtain the best possible outcomes. To effectively conduct a HACCP study, it must be overseen by a supervisor who has the most comprehensive understanding of HACCP concepts (Wallace et al., 2012).

### 2.3.4.2 Stage 2: Describe Product/ Process

This phase encompasses information regarding both the products and the processes, and serves as a helpful guide for food safety and control practitioners who are members of the team. A HACCP system acts as an ideal introductory tool for new employees, as well as a great overview for auditors and inspectors from outside of the company or a third party.

In addition to product/process description, the following details should be supplied:

- Include the main ingredient groups to be employed in processing as well as the "work-in- progress" (WIP) inputs for those processing modules;
- The fundamental procedures and the techniques for handling and preparing materials;
- The manufacturing environment and equipment design;
- If known, the type of hazard is considered;
- Critical control measures incorporated into the framework, procedures, and requirements; if applicable to the scope of the study, packaging/wrapping;
- Product design characteristics that promote safety.

To facilitate in the production of process flow diagrams, it is common practice in food service operations to group all menu/food items into similar process categories at this point.

### 2.3.4.3 Stage 3: Identify Intended Use

It encompasses product abuse, such as inappropriate storage conditions or product usage in aspects other than those originally intended, such as raw cookie dough consumption. Distinct user groups, for example, the elderly, young children, and those with breached immune systems, may be more susceptible to potential hazards. It cannot be stressed enough how important it is for all products to be safe for all consumers. Generally, information about the intended use and consumer group of a product or process is included in the record of its description (from step 2). In order to ensure that the food product's safe handling, storage, and preparation are made clear to the customer, you will have to alert the consumer about the product's intended usage and possible misuse.

### 2.3.4.4 Stage 4: Construct Process Flow Diagrams

Flowchart is used to display the order of tasks that occur when processing procedures. Process flow diagrams are typically used to document processes and for hazard analysis, but in addition, it can inform the subsequent requirements development. To create a flow chart, the procedure must be broken down into a sequence of steps. The term "step" in HACCP pertains not only to prominent processing activities and even to all phases of the product's life, such as raw material receipt and storage. To ensure that the flowchart percolates and resembles how the product is actually created, as well as ensuring that it includes enough detail to help in understanding the manufacturing process and the completion of a full hazard analysis, the flowchart must be comprehensive. HACCP must be concerned with process equipment instead of the processing activity, especially in a commercial food processing environment. To employ a simple flowchart that displays ingredients or groups of ingredients climbing the page to the endpoint, where the finished product(s) are situated, the most widely used HACCP flowchart is frequently used.

### 2.3.4.5 Stage 5: Confirm Accuracy of Process Flow Diagrams

Hazard analysis flowchart verification is especially important since the process flow diagram will be utilized to structure the hazard analysis. Process monitoring verifies the diagrams that are processed to ensure that no variances exist across shifts, including for those that are being processed during separate shifts. Typically, HACCP team or production people are responsible for doing this activity. The hazard analysis should commence once the process flow diagram has been signed off and dated as viable.

### 2.3.4.6 Stage 6: Conduct a Hazard Analysis

Each process step is examined defined one by one by the HACCP team.. It makes a list of any potential hazards that may occur before analyzing to identify significant hazards and appropriate control measures. According to Mortimore and Wallace (2013), allowing HACCP teams to document essential components of prospective risk assessments, reasoning, and decision-making regarding relevance and appropriate control actions, the usage of Hazard Analysis Charts aids in the structuring of the hazard analysis process. Control measures can be determined by evaluating existing controls, but it is critical to determine whether they are sufficient or require additional control.

### 2.3.4.7 Stage 7: Determine Critical Control Points

At critical control points (CCPs), which are points where the most important dangers must be controlled, the vital safety functions must be carried out. HACCP teams often employ the Codex CCP decision tree in their day-to-day operations. For future reference during hazard analysis, it is incredibly beneficial to record the team's discussions and justifications for decisions, which are typically accomplished using a CCP decision record.

### 2.3.4.8 Stage 8: Establish Critical Limits for Each CCP

In order to implement HACCP Principle 3, it is necessary to set firm safety requirements for each CCP, known as critical limits. Products that result from processes operating outside of their critical limits may be hazardous. Critical limits are crucial considerations (never ranges) that denote the demarcation line between "safe" and "potentially unsafe". All CCPs must have quantifiable critical limits, but while operational limits are defined to be "tighter" parameters than those required for safety, setting them at "tighter" parameters than safety requirements ensures that process management parameters, such as alert thresholds, will have a buffer zone, providing an area of intervention before the process crosses a critical limit.

### 2.3.4.9 Stage 9: Establish a Monitoring System for Each CCP

When the critical limits are implemented, a surveillance model is intended to monitor the CCPs continuously. This must demonstrate the effectiveness of the CCPs.

Monitoring: It is the process of executing a predetermined sequential order of measurements or observations of control parameters to determine whether a CCP is in control. Thus every surveillance action should have a designated individual responsible for carrying out the monitoring task, recording the results, and initiating any appropriate measures. Workers on production lines who are involved in the operations where the CCPs are installed are used by manufacturers to monitor the overall manufacturing process. If everything's connected to the network, there should be continuous monitoring systems and alert and response systems linked to each other.

### 2.3.4.10 Stage 10: Establish Corrective Actions

Inaccurate readings disclose deviations from a stated critical limit, and corrective action must be performed. The HACCP team must identify corrective action methods and accountability during the HACCP study so that the suitable team members can enact them in the event of a deviation. Specialized interventions are necessary to safely and promptly deal with the potentially dangerous product and restore the process. In order to successfully validate and invalidate the proposed corrective action plan, the latter must be challenged as the consumer's last line of defense from receiving potentially unsafe products if a CCP fails.

### 2.3.4.11 Stage 11: Establish Verification Procedures

Two distinct types of approval are necessary: validation and verification. Verification entails developing a method to ensure that the HACCP system is capable of and operates appropriately. These activities are distinct and distinguishable from one another.

Validation – Confirming the components of the HACCP system's effectiveness.

Verification – In order to create HACCP compliance, methodologies, processes, tests, and other evaluations are used in conjunction with monitoring to achieve results.

### 2.3.4.12 Stage 12: Establish Documentation and Record-Keeping

A critical component of the documentation will be the HACCP plan, which will detail the CCPs and their management procedures. It demonstrates to external auditors the legitimacy of the approach and decisions.

Thus, practical HACCP requires the maintenance and archiving of HACCP records. While paper archives are still used to store records, an increasing number of businesses are adopting computerized record-keeping systems.

## 2.4 CONCLUSION

Contamination of food during production and processing is diversifying; as a result, control measures should be strengthened and implemented. While food processing methods can help control the level of contaminants to a certain degree, new particulates may also be generated during the manufacturing process due to endogenous factors in foods. Whereas research efforts to identify technologies for detecting food contaminants have increased, there is a dearth of information on the degradation process of contaminants and the toxicity of their degradation products. There is a dearth of standard operating procedures for food transportation, owing to the risks introduced by new logistics. Ultimately, novel compounds are introduced into foods or packaging materials, which may cause a hazard to public health. Such observations emphasize the importance of regulatory safeguards and detection techniques throughout the food manufacturing process to ensure food safety. Continuous reinforcement of hygiene messages in the workplace is critical to sustaining intended food handling practices. Additionally, food hygiene can be improved by providing a physical and social environment conducive to appropriate food handling behaviors. Training sessions that are strongly related to this type of environment would be more suitable than food hygiene courses in settings unrelated to the place of work and rely exclusively on knowledge-based evaluation methods.

## REFERENCES

Balinova A, Mladenova R, and Obretenchev D. 2006. Effect of grain storage and processing on chlorpyrifos-methyl and pirimiphosmethyl residues in post-harvest-treated wheat with regard to baby food safety requirements. *Food Addit Contam* 23: 391–397.

Barendsz AW. 1998. Food safety and total quality management. *Food control* 9: 163–170.

Brambilla G, and Testa C. 2014. Food safety/food security aspects related to the environmental release of pharmaceuticals. *Chemosphere* 115: 81–87.

British Retail Consortium. 2011. Global Standard for Food Safety.

Bryan FL, and World Health Organization. 1992. Hazard analysis critical control point evaluations: a guide to identifying hazards and assessing risks associated with food preparation and storage. Geneva: World Health Organization.

Chung SW, and Chen BL. 2011. Determination of organochlorine pesticide residues in fatty foods: A critical review on the analytical methods and their testing capabilities. *J Chromatogr A* 1218: 5555–5567.

Codex Commission CA. 1997. Hazard analysis and critical control point (HACCP) system and guidelines for its application. *Annex to CAC/RCP* 3: 1–1969.

Da Cruz AG, Cenci SA, and Maia MC. 2006. Quality assurance requirements in produce processing. *Trends Food Sci Technol* 17: 406–411.

Da Cunha DT, Cipullo MAT, Stedefeldt E, and de Rosso VV. 2015. Food safety knowledge and training participation are associated with lower stress and anxiety levels of Brazilian food handlers. *Food Control* 50: 684–689.

Dias MAC, Sant'Ana AS, Cruz AG, José de Assis FF, de Oliveira CAF, and Bona E. 2012. On the implementation of good manufacturing practices in a small processing unity of mozzarella cheese in Brazil. *Food Control* 24: 199–205.

Djekic I, Sanjuan N, Clemente G, Jambrak AR, Djukic- Vukovic A, Brodnjak UV, Pop E, Thomopoulos R, and Tonda A. 2018. Review on environmental models in the food chain – Current status and future perspectives. *J Clean Prod* 176: 1012–1025.

Ehlert KA, Beumer CWE, and Groot MCE. 2008. Migration of bisphenol A into water from polycarbonate baby bottles during microwave heating. *Food Addit Contam* 25: 904–910.

Estevez M. 2015. Oxidative damage to poultry: From farm to fork. *Poult Sci* 94: 1368–1378.

Feglo PK, Frimpong EH, and Essel-Ahun M. 2004. Salmonellae carrier status of food vendors in Kumasi, Ghana. *East Afr Med J* 81: 358–361.

Feglo P, and Sakyi K. 2012. Bacterial contamination of street vending food in Kumasi, Ghana. *J Med Biomed Sci* 1: 1–8.

Food Standards Agency. 2009. Understanding the Influence of Physical and Chemical Properties on Stickiness, Collapse and Open Texture. Project Code N09026. London: FSA.

Gilling SJ, Taylor EA, Kane K, and Taylor JZ. 2001. Successful hazard analysis critical control point implementation in the United Kingdom: understanding the barriers through the use of a behavioral adherence model. *J Food Prot* 64: 710–715.

Groh KJ, Geueke B, and Muncke J. 2017. Food contact materials and gut health: Implications for toxicity assessment and relevance of high molecular weight migrants. *Food Chem Toxicol* 109: 1–18.

Guerrini A. 2005. Beasts of the earth: Animals, humans, and disease. *Emerg Infect Dis* 11: 1162.

Huang C, Zhu J, Chen L, Li L, and Li X. 2014. Structural changes and plasticizer migration of starch-based food packaging material contacting with milk during microwave heating. *Food Control* 36: 55–62.

Jiao X, Zeng J, Yating R, Wang L, Wang J, and Zhang S. 2017. Migration of phthalate esters from packaging and container materials to yellow rice wine. *Liquor-Making Sci Technol* 3: 34–37.

Jing HE, Cheng N, and Wentao XU. 2015. Development of New Rapid Screening Technologies for Microbes in Food. *Food Sci* 36: 288–293.

Kamboj S, Gupta N, Bandral JD, Gandotra G, and Anjum N. (2020). Food safety and hygiene: A review. *Int J Chem Stud* 8: 358–368.

Kobayashi M, Otsuka K, Tamura Y. 2011. Study on rapid analysis method of pesticide contamination in processed foods by GC-MS and GC-FPD. *Food Hygiene Saf Sci* 52: 226–236.

Lalah JO, and Wandiga SO. 2002. The effect of boiling on the removal of persistent malathion residues from stored grains. *J Stored Prod Res* 38: 1–10.

Li X, Huo L, and Liu C. 2018. Study on phthalates plasticizers detection of food paper packaging. *Appl Sci Graphic Commun Packag* 477: 597–601.

Liu X. 2016. Risk analysis and pollution control of phthalate esters in edible vegetable oil. *Modern Food* 16: 94–97.

Mortimore S, and Wallace C. 2013. Preparation and planning to achieve effective food safety management. In *HACCP*. Boston, MA: Springer, pp. 37–66.

Nerín C, Acosta D, and Rubio C. 2002. Potential migration release of volatile compounds from plastic containers destined for food use in microwave ovens. *Food Addit Contam* 19: 594–601.

Nerín C, Canellas E, Romero J, and Rodriguez Á. 2007. A clever strategy for permeability studies of methyl bromide and some organic compounds through high-barrier plastic films. *Int J Environ Analyt Chem* 87: 863–874.

Nerín C, Fernández C, Domeño C, and Salafranca J. 2003. Determination of potential migrants in polycarbonate containers used for microwave ovens by high-performance liquid chromatography with ultraviolet and fluorescence detection. *J Agric Food Chem* 51: 5647–5653.

Patriarca A, and Fernandez Pinto V. 2017. Prevalence of mycotoxins in foods and decontamination. *Curr Opin Food Sci* 14: 50–60.

Rasmusssen RR, Poulsen ME, and Hansen HC. 2003. Distribution of multiple pesticide residues in apple segments after home processing. Food Addit Contam 20: 1044–1063.

Shu- Juan JI, Liu SY, Zhuo YU, and Chang N. 2014. Pattern of aluminum migration from aluminum foil used for food packaging in food simulation. *J Shenyang Agric Uni* 45: 42–46.

Soyer A, Ozalp B, Dalm U, and Bilgin V. 2010. Effects of freezing s temperature and duration of frozen storage on lipid and protein oxidation in chicken meat. *Food Chem* 120: 1025–1030.

Uygun U, Senoz B, and Koksel H. 2008. Dissipation of organophosphorus pesticides in wheat during pasta processing. *Food Chem* 109: 355–360.

Uygun U, Senoz B, Ozturk S, and Koksel H. 2009. Degradation of € organophosphorus pesticides in wheat during cookie processing. *Food Chem* 117: 261–264.

Wallace CA, Holyoak L, Powell SC, and Dykes FC. 2012. Re-thinking the HACCP team: An investigation into HACCP team knowledge and decision-making for successful HACCP development. *Food Res Int* 47: 236–245.

Wei Z, Sheng-Shuang LI, Guo-Qiang YN. 2017. Application of enzyme in preservation of processed food. *China Fruit Vegetable* 1: 7–9.

WHO, HACCP, W. 1997. Introducing the Hazard Analysis and Critical Control Point System. *FOOD SA ISSUES. WHO/FSF/FoS*, 97(2): 1–21.

Wu S. 2017. Preparation of canned apple juice using glutathione as an enzymatic and non-enzymatic browning inhibitor. *J Food Process Preserv* 41: e12750.

Xue MG, Tian LY, and Wang SF. 2016a. Study on the migration of organic contaminants from paper packaging to solid food. *Packag Food Machine* 3 :14–18.

Xue MG, Wang SF, and Lin CY. 2016b. Migration of organic contaminants from the surface of food packaging paper to food. *Packag Eng* 1 :29–34.

Zhang W, Xiao S, Lee EJ, and Ahn DU. 2011. Effects of dietary oxidation on the quality of broiler breast meat. *Anim Ind Rep* 1: 657.

Zukowska J, Biziuk M. 2008. Methodological evaluation of method for dietary heavy metal intake. *J Food Sci* 73: 21–29.

CHAPTER 3

# Green Practices Initiative as a Sustainable Aspect for Food Industry

**Deepika Puri[1] and Bharat Kapoor[2*]**
[1]Chitkara College of Hospitality Management, Chitkara University, Punjab, India
[2]Department of Hotel Management and Tourism,
Guru Nanak Dev University, Amritsar (Punjab), India
*Corresponding author email: bharatkapoor22@gmail.com

## CONTENTS

| | | |
|---|---|---|
| 3.1 | Introduction | 33 |
| 3.2 | Water Conservation | 35 |
| 3.3 | Energy Consumption | 35 |
| 3.4 | Green Study in the Hospitality Industry | 36 |
| | 3.4.1 Green Customers | 37 |
| | 3.4.2 Waste Management | 37 |
| 3.5 | Theoretical Ideology/Mindset of Green Practices | 38 |
| 3.6 | Sustainable Food Systems | 38 |
| 3.7 | Sustainable Food Sources | 39 |
| 3.8 | Sustainable Food Processing | 39 |
| 3.9 | Green Practices in Restaurants | 40 |
| | 3.9.1 Recycling and Composting | 40 |
| | 3.9.2 Energy and Water-Efficient Equipment | 40 |
| | 3.9.3 Servingware and Packaging | 41 |
| | 3.9.4 Menu Sustainability | 41 |
| 3.10 | Conclusions | 41 |
| References | | 43 |

### 3.1 INTRODUCTION

As more customers grasp the gravity of environmental issues, consumer choices are becoming more environmentally conscious, as they select environmentally friendly products and services (Han et al., 2010). To address the growing demand for "green" products and services, marketers across all sectors invest heavily in producing and promoting environmentally friendly items. The emphasis on being environmentally friendly has compelled the restaurant business to alter its offerings in order to satisfy the changing demands of its clients. Restaurant operators are integrating Green Practice (GP) on their premises by becoming members of green groups like the Green Restaurant Association (GRA).

The influence of business image on customer behavior is well known in marketing. According to the researchers, a positive corporate image assists businesses in establishing and maintaining loyal customer relationships (Robertson and Anderson, 1993; Andreassen and Lindestad, 1998; Nguyen et al., 2001). Furthermore, the major reasons why the firms are adopting eco and green practices is to improve the public image in the customer group and reputation of the hotels. According to a new LLP study, around two-thirds, that is 66% of the CFOs of the top 100 biggest retailers, indicated that improving company image is the most important motivator component for companies, especially hotels, to pursue eco-friendly and green practices, with 54% indicating "image among consumers" and around 13% indicating "image among shareholders" (Environmental Leader, 2007). This suggested that the hotel industry experts acknowledged the relevance of eco and green practices as one of the components contributing to the company's image, and that they thought the company's image can be enhanced by implementing green practices, which will contribute to customer loyalty in the long term (Ryu et al., 2008).

Human existence is wreaking havoc on the environment on a daily basis, and restaurants are a contributing factor (Choi and Parsa, 2007). The restaurants are now pushing and adopting numerous sustainable measures to address this issue (Choi and Parsa, 2007). The American public are coming to see environmental sustainability as a lifestyle choice instead of a problem, increasing the growing need for restaurants to include sustainability programs into their operations (Goodland, 1996). As the restaurants generate a lot of garbage, use a lot of energy, and deplete the use of natural resources, environmental sustainability is considered as an important practice in the hotel and restaurant industry in the United States (Barclay, 2012).

Restaurants contribute to environmental degradation through their operations, building, and design. Restaurant trash comprises food, paper and discarded paperboards, waste plastics, metals, and glass pieces (U.S. Environmental Protection Agency, 2010). The largest wastage among all these is food and is a contributor to landfills and various incinerators in the United States, accounting for 15% of all rubbish generated (Barclay, 2012). In 2011, about 36 million tons of food waste was generated, with just 4% of that total composted and not ending up in landfills or incinerators (U.S. Environmental Protection Agency, 2012).

Restaurants are a significant component of commercial and business buildings in the United States, consuming more than three times the total energy per square foot as compared to other commercial buildings (U.S. Energy Information Administration, 2013). The extended hours of operation, massive amounts of equipment, and demand all contribute to a high level of consumption in restaurants; nevertheless, the majority of this energy consumption is typically inefficient (Sustainable Foodservice

Consulting, 2013). Reducing the wastage and energy consumption in the hotels and restaurants is a critical issue that the hospitality sector must solve since restaurants contribute to the challenges of depleting and deteriorating the Earth's resources.

The hospitality industry has discovered several methods to implement environmentally and ecofriendly programs, but changing the behavior and thinking of employees remains an obstacle in the hospitality business (Checkley-Layton, 1997). The employees need to be willing to pursue the initiative policies and sincerely believe in the ideals of environmentalism in order to actually make a difference (Sirota et al., 2005). The effort success is dependent on staff response, and without employee participation, the restaurants' sustainability initiatives would fail (Govindarajulu and Daily, 2004).

## 3.2 WATER CONSERVATION

According to Shanklin (1993), the hospitality and tourist industries should be concerned about the availability and quality of clean water. Water supplies are quickly depleting, causing water stress as a result of increased demand and depletion of nonrenewable fossil water resources, contamination of water related sources, and diminishing levels of precipitation (Gossling et al., 2012). The hotel industry contributes to this issue by accounting a total for 15% of total water usage in commercial establishments in the United States (US Environmental Protection Agency, 2012). According to Hankinson (1992), restaurants will need to divide the water sources in the future:

1) "pure" water for food preparation and drinks and
2) "non-potable" water employing a treatment for ware washing and other cleaning applications.

According to the United States Environmental Protection Agency (2012a, 2012b), kitchens use the most water in restaurants, followed by bathrooms. Brodsky (2005) suggests a few water-saving measures for kitchens, including: 1) adding a low-flow pre-rinse spray valve, 2) using an Energy Star certified steam cooker, and 3) employing other Energy Star rated commercial equipment.

## 3.3 ENERGY CONSUMPTION

High energy expenditure in restaurants is mostly caused by inefficient food cooking, holding, and storage equipment. At an 8% profit margin, even a $1 reduction in energy expenses may increase revenues by $12.50 (Sustainable Foodservice Consulting, 2013). Restaurant energy usage may be classified as refrigerators, preparing food and cooking, computer equipment, sanitation, computers, ventilation, heating, circulation, lighting, and other, according to the Energy Information Administration (2013). According to Sustainable Foodservice Consulting, 2013, preparing food and cooking account for around 24% of total energy usage, followed by other processes like heating, sanitation, and refrigeration which account for 16.7%, 16%, and 15%, respectively.

Several factors impact energy usage in restaurants, including: 1) operating hours, 2) methods of operation, 3) quantity and the kind of appliances, 4) various conditioned space seats, 5) customer traffic patterns, 6) climatic zone, 7) walk-in refrigerator type, and 8) outside lighting (Hedrick, Smith, and Field, 2011). Energy-saving tips include: 1) upgrading equipment to Energy Star qualified appliances; 2) performing various equipment activities like regular cleaning and maintenance; 3) making efficient and prompt repairs; 4) implementing energy efficiently including such occupancy sensors and LED light bulbs; and 5) implementing equipment start-up and completely closed schedules and operations (EnergyStar, 2013).

## 3.4 GREEN STUDY IN THE HOSPITALITY INDUSTRY

Choi and Parsa (2007) stressed the willingness of managers to get engaged in environmentally friendly initiatives. They explored the association between psychological characteristics such as restaurant managers' views, interests, and engagement in green practices and the readiness to charge for green practices. Unlike the other majority of green studies, this one focused on hotel managers' attitudes toward environmental behavior. Furthermore, their study provided a fresh viewpoint on how pricing decisions connected to green practices may be defined by the level of the psychological elements of managers. The findings demonstrated that managers' preferences and participation in such behaviors influenced their proclivity to charge higher fees for engaging in socially responsible activity. Managers' opinions about general practitioners, on the other hand, had little or negligible impact on managers' willingness to charge for completing a socially acceptable practice. This research establishes a solid framework of concept for green and ecofriendly restaurant operations. They insisted that green practices were developed by keeping in mind three important perspectives: health, the environment, and social concern.

Gustin et al. (1996) evaluated consumers' intent to stay at the hotel depending on the company's environmental measures. They used a modified and verified environmental behavioral model for measurement (Hines et al., 1987) to assess hotel guests' intent to purchase a stay at a green hotel, which includes knowledge of the users about environmental concerns, attitudes toward environmental strategies, and perceived self-efficacy. Their findings revealed that three different components of the environmental behavioral model had a favorable connection with purchasing intent. The study was unusual in that it attempted to uncover green practices that may elicit customers' behavioral intentions. This was the first study to look at what customers felt about green practices and their expectations of them in the hotel business.

### 3.4.1 Green Customers

Han et al. (2009) focused on people becoming more environmentally concerned and eager to acquire environmentally friendly products and services as they become more aware of the gravity of environmental concerns. As per Ryan (2006), Americans are growing increasingly worried about the environment and eco-friendly products. The number of Americans, over the last two years who are concerned about the environment has risen from 62% in 2004 to 77% in 2006. Approximately 80% of Americans now buy green goods on a daily or occasional basis. Only 12% of Americans are real greens, while the other 68% are mild greens (Hanas, 2007).

Straughan and Roberts (1999) stressed the relevance of psychographic measures in identifying green clients in their research into environmental segmentation. Psychographic characteristics, according to the researchers, give a more robust and relevant profile of the green intake than the demographic data. According to their findings, the most important predictor of environmentally conscious consumer behavior was perceived customer effectiveness (PCE) in solving environmental problems (ECCB). Consumer attitudes or perceptions that "individuals may favorably affect the outcome of such situations" are referred to as perceived customer effectiveness (PCE) (Straughan and Roberts, 1999). It assesses the extent to which a client can help save the environment. This component is often regarded as one of the most essential factors influencing green consumers' behavior intentions. Customers that suggest a high degree of PCE have higher levels of green purchases, according to green research (Chan and Lau, 2000; Gilg et al., 2005; do Paco et al., 2009).

Restaurants may employ the famous Reduce, Reuse, Recycle program to control trash, including food waste, plastics, glass, and paper/paperboard. The Environmental Protection Agency defines food waste as "uneaten food and food preparation wastes from homes and commercial businesses such as grocery shops, restaurants, and produce markets, institutional cafeterias and kitchens, and industrial sources such as employee lunchrooms" (U.S. Environmental Protection Agency, 2011).

### 3.4.2 Waste Management

Composting, recycling and decreasing or eliminating waste, particularly food waste, are all part of waste management systems. Recycling initiatives have a significant influence on trash management (Wilson et al., 2006). There are two types of recycling programs applicable to the restaurant industry: one with a full-time recycling coordinator role directing recycling operations and another that adds recycling tasks to present staff (Lounsbury, 2011). According to Snarr and Pezza (2000), restaurants may decrease waste by purchasing recycled materials, setting purchasing criteria, giving unsold things to local charities and food banks, reusing items rather than dumping them, making food to order, and ensuring correct food storage protocols are in place.

## 3.5 THEORETICAL IDEOLOGY/MINDSET OF GREEN PRACTICES

According to Miles et al. (2000), the following are two ideas that briefly explain the reason organizations invest in generating the superior environmental performance:

(1) The "slack resources" view (Graves and Waddock, 1994) and
(2) The "excellent management" competitive advantage perspective (Russo and Fouts, 1997).

The first idea, "slack resources," contends that a corporation with adequate assets will spend discretionary resources to socially acceptable actions such as environmental improvements. This investment is intended to build and strengthen competitive advantage through reputation and cost savings (Miles and Russell, 1997). In other words, by delivering exceptional green performance, the firm hopes to improve its image and reputation, which might lead to more successful results in the future. According to good management theory, organizations with innovative management seek out developing sources of competitive advantage, such as new environmental practices, to better please consumers.

Hotel managers are concerned about achieving superior and better environmental performance, which reflects the customers' recognition of the company's performance of green image through the company's green practices (performance), because it has been believed that achieving such performance in public will provide a distinct advantage that will increase their competitive power.

According to the above given two theories, regardless of what the financial situation, the company or the group of people, or the management methods, the fundamental reason why organizations participate in conducting and growing better environmental performance in order to improve the company's image through green practices and, eventually, to obtain a competitive edge. According to a recent research, two-thirds of the top hundred largest retailers' CFOs feel that the most important reason for a company to embrace ecofriendly practices is to improve its image, with 54% t citing "image among customers" and 13% suggesting "image among shareholders." The poll also revealed that industry experts regarded the importance of eco-friendly activities as one of the variables contributing to the company's image as one of the factors contributing to the company's image (Environmental Leader, 2007). Industry experts believe that by using environmentally friendly procedures, the company's image may be improved, which can lead to long-term client loyalty (Ryu et al., 2008).

## 3.6 SUSTAINABLE FOOD SYSTEMS

Sustainable food systems (SFS) are critical to maintaining long-term development and food and nutritional security for all people. The first book (Knorr, 1983) dealing with sustainable food systems states in its preface that "efforts have remained limited in scope, and they often provide only sporadic responses to far-reaching problems."

Unfortunately, these assertions may still be accurate today. Pimentel and Pimentel (2007) analyzed fossil energy requirements vs. US daily food energy consumption for various diets and found that non-vegetarian diets require ten times the amount of fossil energy as vegetarian diets.

A selection of unsolved food science—related issues and recommended techniques for solutions in developing nations from the US National Academy of Science—National Research Council in 1974 (Brown, 1975) seems like a very recent paper, indicating a lack of development in the field.

Gustafson et al. (2016) suggested the seven sustainability indicators. Chen et al. (2009), who evaluated sustainability across seven areas, revised them (nutrition, environment, food affordability and availability, sociocultural well-being, resilience, food safety, and waste). Sustainability has also been investigated through the lens of product and nutritional demands, while integrating interlinkages between various food and supply chains, taking into account sourcing, processing, and transportation, environmental aspects (e.g., land use, climate change, fossil fuel depletion, etc.), and costs, with Life Cycle Assessment serving as a foundation.

## 3.7 SUSTAINABLE FOOD SOURCES

There is a need to establish renewable and sustainable food sources. Meeting the rising worldwide demand for proteins in a sustainable manner is a serious problem. Approximately one-third of all food produced is thrown away. The edible fraction of food, which is currently not reaching the food supply, is an emerging source of novel components for the food processing sector. To make edible biomass a new supply of raw materials for food processing, steps should be done to recover and repurpose edible biomass that is now discarded. Reducing food loss and waste, along with other agricultural productivity-boosting techniques, should be part of the solution for sustainable food systems. To determine the prioritization of useful actions, an assessment of both environmental and economic consequences for adopting measures to decrease food waste is required.

## 3.8 SUSTAINABLE FOOD PROCESSING

Thermal processing and the exploitation of natural renewable resources have historically been the driving forces behind developments in food preparation. These include the use of low or high temperatures for processing and preservation, which a date back to William Cullen's creation of the first artificial cold process in the 1750s and has progressed to contemporary energy efficient refrigeration and freezing systems. Similarly, Nicholas Appert pioneered the use of heat for food preservation in the early 1800s in response to Napoleon Bonaparte's need to feed the French army, which has now developed into contemporary energy-efficient thermal-processing technology.

Green practices that demonstrate social responsibility have long been seen as an important component in determining business reputation, as well as a key component of corporate image (Schwaiger, 2004). Furthermore, marketing studies show that these practices have a substantial impact on the appraisal of a company's image, reputation, and even customer loyalty (Dutta et al., 2008). In theory, green practices might be a minor component of a company's overall image. However, given the contemporary social atmosphere in which people are concerned about the environment and want products and services that are less detrimental to the environment, there has been a deliberate attempt to go green. Companies have also been forced to become more aware of the present situation of the environment. Because of the heightened social sensitivity to this topic, a company's image can be severely hurt by a perceived lack of interest in environmental matters.

## 3.9 GREEN PRACTICES IN RESTAURANTS

There are various green restaurant groups that offer online tools to assist restauranteurs to adopt green practices. Based on a thorough review of the literature, this study identified green practices that may be implemented in the restaurant industry.

### 3.9.1 Recycling and Composting

Restaurants generate a large amount of recyclable rubbish. Glass, plastic, metal, cardboard, and aluminum are among them. Composting food waste helps to minimize waste while also improving soil quality. These are some examples of feasible green restaurant practices for recycling and composting:

- Recycling the paper, plastic material, cardboards, glasses, and aluminum at the back of the house
- Providing recycling bins in shop specially self-service restaurant setting
- Conducting various food waste composting programs

### 3.9.2 Energy and Water-Efficient Equipment

Energy and water saving technology may be used in a variety of locations within a restaurant, including the kitchen, restroom and dining area. Here are several examples:

- Use of flow restrictors on faucets, low-flow toilets, and water-less urinals
- Motion detectors usage for lights in the restroom
- Only serve customers water upon request
- Replacing incandescent light bulbs with longer lasting CFL light bulbs or LED
- Replace exit lights with LEDs or more usage of LEDs
- Use of a system which monitors and controls comfortable temperatures efficiently with the HVAC (Heating, Ventilating and Air Conditioning) system supplies.

Non-toxic cleaning supplies are safe for the environment and people in the following examples:

- Use of environmentally friendly cleaners for dishes, and linen
- Use of environmentally friendly cleaners for tables and floors.

### 3.9.3 Servingware and Packaging

Recycle service products are manufactured from post-consumer garbage. These products can help to decrease trash. Also, they can save natural resources, such as trees.

- Instead of using Styrofoam, use biodegradable (paper) or recyclable take-out containers.

### 3.9.4 Menu Sustainability

Organic food is grown using non-toxic pesticides and fertilizers, and it is not genetically engineered. Locally grown foods help to lessen the amount of air pollution caused by transportation that requires fossil fuels. As a result, restaurant managers must ensure that they:

- Offer organic food on the menu
- Offer fish and seafood harvested sustainably and free of harmful pollutants
- Avoid genetically modified foods.

Green practices derived from these sources were created for restaurant owners. Other techniques for the back of the home were not considered, such as employing energy-efficient lighting in storage and kitchens. This research looked at green practices that customers are exposed to.

## 3.10 CONCLUSIONS

Restaurant managers should work on upgrading the restaurant image to demonstrate that the firm cares about the environment by undertaking GP to elicit consumer behavioral intention. This objective may be met by implementing practical and visible green initiatives in restaurant locations. Cups, napkins, or cup sleeves, for example, are the most accessible tools or utensils needed by consumers in a coffee shop context. Employing glasses or cups instead of throwaway cups, or using recyclable napkins or glasses or cup sleeves that appeal to consumers in a tangible way would assist management in improving the image of the eateries. Furthermore, increasing customer engagement in green activities may be a major approach for improving the restaurant's green image, which in turn enhances consumers' behavioral intentions toward the restaurant in the long run. According to the survey findings, "providing recycling bins for cups and sleeves at the business" placed highest in the evaluation of green measures in coffee shops.

As a result, managers should build such green features in order to provide direct experience chances to restaurant consumers. Managers can also teach their personnel to educate consumers on the recycling opportunities available in the business. Managers may help to enhance the restaurant's image by posting signs informing guests about how they can engage in green practices to keep the environment clean.

According to the findings of this study, customers' perceived the green image of the Green marketing from businesses, rather than customers, may have a significant impact on restaurants. Restaurant patrons' perceptions of green practices. In other words, even if the firm executes good green practices, customers may under-perceive the restaurant's green image, and customers' perceived green image of the restaurant can be developed through the companies' green advertising, regardless of the companies' green practices performance. This implies that by performing green advertising, restauranteurs may effectively inspire a customer's intention to engage in environmentally friendly activity. Furthermore, managers should consider the firm's resources and optimize the usefulness of achievable green practices that provide their clients in each segment the most favorable behavioral intent toward the organization.

One method for achieving this goal is for restaurant owners to perform a customer survey to determine which green segments the business is working with. The top green methods effective for the individual restaurant should be implemented based on the green segments. For example, a restaurant with limited financial resources that wants to implement green practices might start with a green practice like "using recyclable take-out containers," especially if the bulk of the customers are in the less green category. They can have a trash recycling system and energy efficient equipment after they have enough money to add more green qualities in the restaurant.

Despite its relevance, this study has numerous drawbacks. It is still debatable whether consumers' perceptions of the restaurant's efficacy in implementing green practices play a significant part in their decision-making process. Other restaurant features, such as food, services, environment, and pricing, which may have a direct impact on consumers' perceived image of the restaurant as well as their behavioral intention, were not included in this study. Future research may include green practices as one of the restaurant qualities and investigate the influence of the green attribute on behavioral intention in conjunction with an examination of other restaurant features.

Furthermore, in order to quantify customers' perceived ecological image of the restaurant, this study focused on a green image determinant, which is consumers' impression of green practices. As indicated in the literature study, additional physical signals such as the restaurant's brand name and décor can also be factors in building a green picture of the restaurant in the minds of guests. As a result, future research should explicitly define other green image factors and investigate the link between those characteristics and consumers' perceptions of the restaurant's green image.

## REFERENCES

Andreassen TW, and Lindestad B. 1998. The effect of corporate image in the formation of customer loyalty. *J Serv Res* 1: 82–92.
Barclay E. 2012. For restaurants, food waste is seen as low priority. *NPR The Salt*.
Brodsky S. 2005. Water conservation crucial to energy savings. *Hotel Motel Manage* 220: 12.
Brown NL, Pariser E. 1975. Food science in developing countries. *Science* 188: 589–593.
Chan RYK, and Lau LBY. 2000. Antecedents of green purchases: A survey in China. *J Consum Mark* 17: 338–357.
Checkley-Layton A. 1997. Special report: strict audits and formal management structures help Canadian Pacific hotels meet their environmental goals. *The Magazines of the Worldwide Hotel Industry* 3: 13–14.
Chen C, Chaudhary A, and Mathys A. 2019. Dietary change scenarios and implications for environmental, nutrition, human health and economic dimensions of food sustainability. *Nutrients* 11: 856.
Choi G, and Parsa HG. 2007. Green practices II: Measuring restaurant managers' psychological attributes and their willingness to charge for the "Green Practices." *J Food Serv Bus Res* 9: 41–63.
do Paco F, Arminda M, Raposo B, and Lino M. 2009. Identifying the green consumer: A segmentation study. *J Target Meas Anal Market* 17: 17–25.
Dutta PN, and Choudhury BS. 2008. A generalisation of contraction principle in metric spaces. *Fixed Point Theory Appl* 1–8.
EnergyStar. 2013. Restaurants. Retrieved from www.energystar.gov/buildings/facilityowners-and-managers/small-biz/restaurants
Environmental Leader. 2007. 83% of largest retailers involved in green practices. Retrieved from www.environmentalleader.com/2007/10/01/83-of-largest-retailers-involvedin-green-practices/
Gilg A, Barr S, and Ford N. 2005. Green consumption or sustainable lifestyles? Identifying the sustainable consumer. *Futures* 37: 481–504.
Gossling S, Peeters P, Hall CM, Ceron JP, Dubois G, Lehmann LV, and Scott D. 2012. Tourism and water use: Supply, demand, and security. An International review. *Tour Manag* 33: 1–15.
Govindarajulu, N., and Daily, B. 2004 Motivating employees for environmental improvement. *Industrial Management and Data Systems* 104(3): 364–372. Retrieved from http://search.proquest.com/docview/23490704?accountid=4117
Graves, S. B., and Waddock, S. A. 1994 Institutional owners and corporate social performance. *The Academy of Management Journal* 37(4): 1034–1046.
Gustafson D, Gutman A, Leet W, Drewnowski A, Fanzo J, Ingram J. 2016 Seven food system metrics of sustainable nutrition security. *Sustainability* 8:196. doi: 10.3390/su8030196
Gustin ME, and Weaver PA. 1996. Are hotels prepared for the environmental consumer?. *Hosp Res J* 20: 1–14.
Han H, Hsu LTJ, and Lee JS. 2009. Empirical investigation of the roles of attitudes toward green behaviors, overall image, gender, and age in hotel customers' eco-friendly decision-making process. *Int J Hosp Manag* 28: 519–528.
Han H, Hsu LTJ, and Sheu C. 2010. Application of the theory of planned behavior to green hotel choice: Testing the effect of environmental friendly activities. *Tour Manag* 31: 325–334.
Hanas J. 2007. A world gone green. *Advertising Age* 78: S-1-S-11.

Hankinson MG. 1992. The greening of food and beverage operations: Management practice and the law. *Hosp Tour Edu* 4: 9–14.

Hedrick R, Smith V, and Field K. 2011. *Restaurant Energy Use Benchmarking Guideline* (No. NREL/SR-5500-50547). National Renewable Energy Lab.(NREL), Golden, CO (United States).

Hines JM, Hungerford HR, and Tomera AN. 1987. Analysis and synthesis of research on responsible environmental behavior: A meta-analysis. *J Environ Educ* 18: 1–8.

Knorr DW. 1983. *Sustainable Food Systems*. Westport, CT: AVI Pub. Co.

Lee MDP, and Lounsbury M. 2011. Domesticating radical rant and rage: An exploration of the consequences of environmental shareholder resolutions on corporate environmental performance. *Bus Soc* 50: 155–188.

Miles MP, and Covin JG. 2000. Environmental marketing: a source of reputational, competitive, and financial advantage. *J Bus Ethics* 23: 299–311.

Miles MP, and Russell GR. 1997. ISO 4000 total quality environmental management: the integration of environmental marketing, Total quality management and corporate environmental policy. *J Qual Manag* 2: 151–168.

Nguyen N, and Leblanc G. 2001. Corporate image and corporate reputation in customers' retention decisions in services. *J Retail Consum Serv* 8: 227–236.

Pimentel D, Pimentel MH. 2007. Food, Energy, and Society. Boca Raton, FL: CRC Press.

Robertson DC, and Anderson E. 1993. Control system and task environment effects on ethical judgment: An exploratory study of industrial salespeople. *Organ Sci* 4: 617–644.

Russo MV, and Fouts PA. 1997. A resource-based perspective on corporate environmental performance and profitability. *Acad Manag J* 40: 534–559.

Ryan B. 2006. Green consumers: a growing market for many local businesses. In *Let's Talk Business Ideas of Expanding Retail and Services in Your Community*. Madison, WI: University of Wisconsin Extension, Draft issue: 123, pp. 1–2.

Ryu K, Han H, and Kim TH. 2008. The relationships among overall quick-casual restaurant image, perceived value, customer satisfaction, and behavioral intentions. *Int J Hosp Manag* 27: 459–469.

Schwaiger M. 2004. Components and parameters of corporate reputation—An empirical study. *Schmalenbach Bus Rev* 56: 46–71.

Shanklin CW. 1993. Ecology Age: Implications for the hospitality and tourism industry. *J Hosp Tour Res* 17.

Sirota D, Mischkind L, and Meltzer M. 2005. 33 beliefs about work and workers'. *The Enthusi-astic Employee*.

Snarr J, and Pezza K. 2000. *Recycling guidebook for the hospitality and restaurant industry*. Washington DC: Information Centre Metropolitan Washington Council of Governments.

Straughan RD, and Roberts JA. 1999. Environmental segmentation alternatives: a look at green consumer behavior in the new millennium. *Journal Consum Market* 16: 558–575.

Sustainable Foodservice Consulting. 2013. Energy conservation.

US Department of Energy, U.S. Energy Information Administration. 2013. Independent statistics & analysis.

US Environmental Protection Agency. 2010. Energy star guide for restaurants: Putting energy into profits (EPA430-R-09-030).

US Environmental Protection Agency. 2011. Glossary of terms: Environmental terms.

US Environmental Protection Agency. 2012a. Non-hazardous waste: Basic information.

US Environmental Protection Agency. 2012b. Saving water in restaurants (EPA-832-F-12-032).

Wilson DC, Velis C, and Cheeseman C. 2006. Role of informal sector recycling in waste management in developing countries. *Habitat Int* 30: 797–808.

CHAPTER 4

# Improving Food Processing Using Various Technologies in the Food Industry

**Monika Thakur**[1] **and Sayeeda Kousar Bhatti**[2*]
[1]Division Botany, Department of Bio-Sciences, Career Point University, Hamirpur (H.P.), India
[2]Department of Botany, Govt Degree College Doda City, Jammu and Kashmir UT, India
*Corresponding author email: sayeedakhan@gmail.com

### CONTENTS

| | | |
|---|---|---|
| 4.1 | Introduction | 45 |
| 4.2 | Emerging Technologies in Improving Food Processing | 47 |
| | 4.2.1 High Hydrostatic Pressure | 47 |
| | 4.2.2 Ultrasound Technology | 49 |
| | 4.2.3 Dielectric Heating | 50 |
| | 4.2.4 Pulsed Light | 51 |
| | 4.2.5 Bacteriocins | 51 |
| 4.3 | Synthetic Biology in the Food Industry | 52 |
| 4.4 | Nanotechnology in Food Processing | 53 |
| 4.5 | Conclusion | 53 |
| References | | 55 |

### 4.1 INTRODUCTION

The food industry's viability increasingly depends on giving attention to product preferences. Reformulating products to include healthier sustainable components, increasing proteins, vitamins, and antioxidants to foods, as well as labelling items as allergen-free, gluten-free, non-GMO, organic, and antibiotic-free are all notable sources. Furthermore, calorie-dense carbohydrates are being phased out or reduced in food, while measures are being done to increase the shelf life of food and prohibit counterfeit items from entering the supply chain (Bhargava et al., 2021). In order to properly execute line production, the manufacturing industry must emphasize the need to include more adaptable equipment based on bioengineering, technology, and robotics, provided the changed conditions of the foodservice industry. This strategy would encourage the pathways and techniques that users seek, resulting in significant changes in the manufacturing and distribution of food (IFI, 2010). The new technology that would undergird the food industry must be modernized in such a manner that it is adaptable, easy to adjust to minimal trade changes in real-time, focused on stringent, long-term cleanliness, and handled by competent technologists, particularly when it comes to bioengineering.

In terms of technology, this leads to faster sensing equipment processing times, self-diagnosis for preventive maintenance, and increased communication and compatibility between processes, ensuring food quality, accountability, and authenticity (Negi, 2012). Manufacturers are looking for information gathering and the Industrial Internet of Things (IIoT) to better manage production schedules, resources, personnel, and maintenance. Many of the strategies and tools that will be capital-intensive solutions, including, in a strategy, the participation of human capital in the development of its capacity technological or financial capital to provide an efficient food experience to each of the communities it serves, would be a concern to perform on the right scale with a growing and demanding base of inhabitants (Nile et al., 2020). This might include the production of genetically modified (GM) crops in inland agriculture, and researchers estimate monitors, algorithms, data fusion, machine perception, and robotics to alleviate several of the roles and productivity constraints of current outdoor agriculture practices within the next ten years.

Smaller, automated warehouses working closer to city centres would be a possibility, with commodities completed by small autonomous electric trucks capable of delivering fresh food to small traders or transforming into outlets (Briones-Labarca et al., 2019). Changes in user preferences, as well as advances in bioscience, biotechnology, and bioengineering, have an impact on where our food is produced. Bioengineering provides a major boost to the production of food bio-products and components with an important nutritional value in novel functional and astute foods, ensuring the long-term viability of existing and emergent food technologies (Condón-Abanto et al., 2016). Food processing is the process of transforming agricultural produce into more edible, shelf-stable, accessible, and useful value-added products that are safe for human consumption through a series of unit activities (Bhargava et al., 2021). Drying, frying, smoking, salting, pickling, soaking, and other traditional processing and preservation techniques are still widely and effectively employed to process and analyze food products. Most traditional food processing procedures rely on the use of heat to reduce the growth of germs and restrict the spread of foodborne illnesses, making the food safe to eat (Negi, 2012). These thermal treatments use a lot of energy and have a low productivity level. They're also time-consuming and require a lot of external inputs (Barba, 2017). Many food ingredients provide a risk of bacterial or viral poisoning, making heat processing unsuitable. Thermally sensitive food products may experience physical, chemical, and microbiological changes as a result of heat treatment, including changes in flavour, colour, and texture. This has necessitated research and development into maximizing the use of existing technologies as well as the development of novel and effective alternative technologies (Misra et al., 2016). High-pressure processing, dielectric heating, pulsed electric field, Bacteriocins, ultrasound, nanotechnology, and synthetic biology are some of the revolutionary food processing techniques.

## 4.2 EMERGING TECHNOLOGIES IN IMPROVING FOOD PROCESSING

Food deteriorates due to physical, chemical, and biological changes that occur after it is harvested. Food preservation is a technique for preventing microbial infection. These microbes, along with enzymes, are the principal agents of change and, as such, should be the emphasis of conservation efforts (Morata, 2010). The following are some of the more recent emerging technologies for preserving food, reducing or eliminating the number of critical pathogenic bacteria in food, and/or extracting bioactive substances helpful in the food industry.

### 4.2.1 High Hydrostatic Pressure

High hydrostatic pressure (HHP) is a technique for physically and chemically altering any chemical molecule contained in food to improve its quality; it is also known as cold pasteurization or pressurization. The food industry is engaged in this technique because of its effectiveness in preserving food, which makes it preferable to traditional thermal techniques (Aguilar et al., 2019), which invariably results in a loss of nutritive value as well as sensory qualities (Aguilar et al., 2019). HHP seems to be the most promising strategy from a commercial perspective (Aguilar et al., 2019), and the one that has demonstrated effectiveness in the inhibition of bacterial spores and enzymes among the alternative (nonthermal) treatments currently known for food preservation (high-intensity electrical pulses, oscillating magnetic fields, high-intensity luminous pulses) (Meyer et al., 2000). The ensuing are some of the benefits of HHP therapy over other non-thermal techniques: – The HHP treatment prevents food from being altered by consistently and abruptly applying pressure to the system, i.e., there are no pressure gradients. HHP treatment, unlike thermal methods, is independent of the volume and shape of the sample, reducing the time necessary to process vast amounts of food (Aguilar et al., 2019). In comparison to standard pasteurization procedures, HHP does not induce the breakdown of thermolabile components like vitamins (due to its low temperature) and does not alter the function and availability of low molecular weight components such as those responsible for the flavour and quality of food (Meyer et al., 2000). Foods treated with high hydrostatic pressure preserve their nutritional and organoleptic features, implying that there is no significant loss of quality as compared to a fresh product. The quality and freshness of HHP processed foods are preserved because of the slight chemical effect induced by the pressure temperature-time parameters utilized, yet the sensory and nutritional qualities remain unchanged. Various studies have shown that products, processed under high pressure, retain the qualities of new products, making it impossible to distinguish them from a fresh untreated product in most circumstances (Aguilar et al., 2019).

The HHP system allows for the destruction of microorganisms that cause the deterioration as well as the decline in quality of various foods; even so, the conditions of the system must be adapted for the different food groups to be processed, so it is recommended that for each type of new product, parameters such as pressurization time and temperature, availability of one or more antimicrobials, and so forth, should be deemed. Pressure rise and fall times in pressurization range from 1 to 5 minutes, depending on the technology (Briones-Labarca et al., 2019). During processing, the food is compressed by 10 to 15% in the range of 300 to 700 MPa, and the heat is raised by adiabatic heating to an estimated ratio of 3°C per 100 MPa of pressure applied in the system (Bhargava et al., 2021). These alterations are only transitory, and after depressurization, the normal conditions are restored. The rapid change in pressure effect microbial deactivation in this procedure, having caused the phase transition of the lipids in the bacterial membrane, and the rupture of ionic bonds, hydrophobic interactions, and the formation of hydrogen bridges, between the molecules, without affecting the covalent bonds, resulting in the emergence of macromolecules as proteins. This has a deleterious impact on other structural and functional aspects of microbial cells, causing them to lose viability (Morata, 2010).

Because pressurization does not affect macromolecules' covalent bonds, it has no effect on vitamins, pigments, or other compounds, preferring their usage in the preservation of these components over traditional heat treatments. Damage to the cell membrane is thought to be the most common cause of cell death, however cell wall, and nucleic acid damage, as well as nucleic acid damage, have also been observed. The structural, physicochemical, and morphological features of pectin treated with high hydrostatic pressure and high-pressure homogenization were investigated by Xie et al. (2008).

According to the research, high-pressure treatment is an effective method for converting pectin from potato peel residues into a thickening or stabilizing agent, but high-pressure homogenization produces a better result. Another study intended to optimize the individual and combined effects of high-pressure operation and solvent polarity (solvent mix) on extraction yield, flavonoid, and lycopene content of tomato pulp (Aguilar et al., 2019). High hydrostatic pressure, the study observed, could be a valuable technique for increasing the extraction as well as the emission of chemicals that could be essential for health (Briones-Labarca et al., 2019). Furthermore, because of its capacity to inactivate harmful and deteriorating microorganisms while minimizing food quality reduction, high hydrostatic pressure treatment is being used as a novel food preservation approach. As a result, the impacts of sublethal (100 MPa) and lethal (N100 MPa) pressures on protein synthesis, structure, and functionality in bacteria are analyzed. High hydrostatic pressure, when combined with specific enzymatic activities, has the possibility to be very useful in biotechnological processes (Gayán et al., 2017).

## 4.2.2 Ultrasound Technology

It's an environmentally friendly, green technology that's successfully improved a variety of procedures in the food industry. It's also a great replacement for a number of heat-based, traditional techniques that degrade the quality of the product. Tenderization and preservation of meat, hardness of fruits and vegetables, improved mixing of doughs, microbial inactivation, homogenization, sterilization, pasteurization, and emulsification all seem to be advantages of ultrasound (Rana and Meena, 2017). Ultrasonication produces a higher-quality product at lower temperatures while also improving heat and mass transfer rates. Ultrasound speeds up the filtration process, extending the life of the filter. It also speeds up the freezing process, resulting in smaller crystal sizes, faster drying, and thawing processes (Carcel et al., 2019).

Ultrasound is a fast processing technology that lowers manufacturing costs. It increases yield and quality and enhances efficiency levels by removing the requirement for process steps. The end product's quality and purity are also improved by enhancing its organoleptic qualities, hardness, and texture. It also promotes the preservation of the product's nutritional quality and prolongs its storage life (Zhang and Abatzoglou, 2020).

While being used maximum intensity, ultrasound generates heat as a result of a temperature increase, which has negative effects on the nutritional and organoleptic aspects of the food. In addition, ultrasound's usefulness against microbial and enzymatic inactivation has not been proven. Therefore, the synergistic impact, in combination with pressure and temperature, could result in inactivation (Bhargava et al., 2021). Furthermore, high-powered ultrasound has the potential to have negative physical and chemical impacts on foods.

Cavitation produces free radicals, which cause lipid oxidation, off-flavors and aromas, protein denaturation, and a decrease in total phenolic content due to ascorbic acid degradation (Condón-Abanto et al., 2016). The use of ultrasound in combination with temperature and pressure causes the creation of free radicals, which catalyze reactions that might disrupt protein structure and therefore impact the texture of the meal. As a result, before using ultrasound, the intensity and synergy must be optimized. Various study examinations have reported the efficacy of ultrasound in the food industry as a substitute, enhancement, as well as modification of several traditional processing procedures (Huang et al., 2020) (Table 4.1).

Therefore, combining ultrasound with other techniques produces greater benefits in terms of the total quality of the product. The optimization of parameters and ground-level research to examine the effect of acoustic stimulation on the quantity of food production are required for the continuous growth of ultrasound for industrial purposes (Cruz et al., 2016). On a bigger scale, studies should investigate the safety features and detrimental impacts of ultrasonography on individuals. Furthermore, the commercialization and industrialization of ultrasound need a tremendous amount of energy, which prevents their use in the food business (Bhargava et al., 2021). As a result, ultrasonic studies should be focused on an application on a large process.

**Table 4.1 Application of Ultrasound in Food Industry**

| Type | Foods | Advantage | Reference |
|---|---|---|---|
| Meat Industry | Chicken, Beef, Pork, Rabbit meat | Enhances tenderness; Improves water dynamics of tissue; Increase in water holding capacity; Color enhancement; pH escalation & antimicrobial effect; Shortens aging period | Caraveo et al. (2015); Wang et al. (2018) |
| Fruits & Vegetables | Fresh and minimally processed fruits and vegetables, Juice, Purees, Edible and refined vegetable oil | Decline of microbial load. (Sterilization & Disinfection); Alterations in color; Enzyme inactivation & Desensitization; Increases drying characteristics.; Purity and quality evaluation of oils | Azam et al. (2020) Chen et al. (2020) |
| Cereal Products | Flour dough, Bakery products (Bread, Biscuits, Crackers, Wafers, Batters (Donuts, pancakes | Assessment of dough and batter assets: texture& rheology, density, volume index. Improved firmness increases texture and colour components, Improves sensory as well as visual aspects | Zhu and Li (2019); Bhargava et al. (2021) |
| Dairy Industry | Milk, Yogurt, Cheese, Ice Cream | Microbial Inactivation, Homogenization, Decrease of fat globules Improvement of organoleptic properties as well as nutritive value, Decrease in time for cheese ripening and fermentation | Guimarães et al. (2019) |
| Emulsions | Mayonnaise, Dressings, Creams, Oil emulsions | Rises in the stability index, emulsion potential, and emulsion activity index | Albano and Nicoletti (2018) |

### 4.2.3 Dielectric Heating

The dielectric heating approach is being used to preserve and improve the quality of foodstuff. Usually employed for commercial purposes, this technology uses a frequency range of 1–300 MHz, with 13.56, 27.12, and 40.68 MHz being the most common. In comparison with other conventional treatment methods, radiofrequency treatment provides considerable benefits like faster, good quality, more uniform heat distribution, and improved energy efficiency for solid and semi-solid foods with low thermal conductivity (Ozturk et al., 2018). Wheat germ, for example, is a desirable by-product of wheat milling, but it is prone to lipid rancidity caused by lipase activity. The dielectric constant and dielectric loss factor of wheat germ increased with rising temperature and moisture content, according to this study. Therefore, the studies of Ling et al. (2018) are valuable in computer simulation and process parameter optimization for wheat germ stabilization using radio frequency heating. Non-uniform heating, on the other hand, is a significant barrier when applying radiofrequency heat treatment in the pasteurization of low moisture food products. This research helped to design an effective RF approach for pasteurizing low moisture materials as an alternative to traditional thermal processing. As a result, it is a revolutionary technique of pasteurizing food powder based on radio frequency heating (Ozturk et al., 2017).

### 4.2.4 Pulsed Light

Pulsed light is a photonic technique that is mostly used in the food technology industry for the inactivation of microorganisms. It includes the use of high-intensity broad-spectrum light pulses, which include UV radiation. The creation of high photon-fluxes is a key element of pulsed light technology (Martinez-López et al., 2019). Cronobacter sakazakii and Salmonella spp., for example, are foodborne infections caused by low moisture diets. Chen et al. (2019) developed an intense pulsed light system as a unique strategy for pasteurizing powdered food. The goal of this study was to use a vibratory-assisted intense pulsed light system to evaluate the microorganism inactivation in various powdered meals and a range of related factors. After numerous passes, the analyses showed that this vibratory-assisted intense pulsed light system may produce increased microbial inactivation. In another study, stimulation with intense pulsed light at a total fluence of 7.40 J/cm2 resulted in a 7 log reduction, demonstrating that intense pulsed light has the ability to accurately neutralize microbial cells. Because the number of vegetative cells grew with the incubation duration, the best results showed that the inactivation efficiency increases after one hour of pre-incubation (Jo et al., 2019).

### 4.2.5 Bacteriocins

Bacteriocins are antibacterial and bioconservative peptides derived from various bacteria strains that serve a vital role in food preservation (Settanni and Corsetti, 2008). They are generated ribosomally and extracellularly and can be utilized in a variety of food systems. Although several bacteriocins have been extracted and characterized, food biopreservation has primarily concentrated on lactic acid bacteria bacteriocins (García et al., 2010). Nysin, diplococine, acidophiline, bulgarican, helveticine, lactaine, and plantaricin seem to be the most significant (Savadogo et al., 2006). It can be found in meat, dairy products, canned goods, fish, vegetables, fruit juices, and beverages such as beer and wine. Its compliance with these products, as well as its mechanism of action, makes it an attractive prospect for usage in food. Pure starter cultures with the ability to produce bacteriocin are being used to make safer fermented fish products in recent years. The genus Lactobacillus, which includes *L. casei*, *L. paracasei*, *L. johnsonii*, *L. plantarum*, and *L. rhamnosus*, generally forms a major part of widely utilized probiotic bacteria in meat (Wang et al., 2015; Avaiyarasi et al., 2016). Nisin is a bacteriocin generated by Lactococcus lactis spp. and is the most analyzed bacteriocin to date, as well as the only bacteriocin used as a food supplement in the world. By adding bacteriocin-like substances (BLIS) synthesized by Enterococcus faecium ES216 with antibacterial activity against Listeria innocua ATCC33090, Salvucci et al. (2019) created active triticale flour films. Triticale flour films activated with these bacteriocins, according to researchers, could be a viable alternative for active food packaging applications.

A *Lysinibacillus* strain isolated from rotten fruits and vegetable waste was found to have an adverse influence against foodborne pathogens in another analysis. A novel bacteriocin of class III was discovered after an investigation of various physiochemical properties and characterization using the 16S rRNA gene. This bacteriocin proved remarkably effective against Bacillus pumilus, a foodborne pathogen (MIC-22 g/mL). According to researchers, *Lysinibacillus* JX402121 isolate can be used to make bacteriocin, which functions as a bio-preservative agent against a variety of food-borne diseases (Ahmad et al., 2019). Bacteriocins are inhibited by proteases, including those of pancreatic and gastric genesis, due to their protein nature. As a result, they are inhibited throughout their passage through the gastrointestinal system, without being absorbed as active chemicals and so relatively safe for consumption (Quintero et al., 2010).

## 4.3 SYNTHETIC BIOLOGY IN THE FOOD INDUSTRY

One of the primary aims of food companies is to ensure the quality and biosecurity standards of their products, as well as the quality and nutrition systems of their products. Producing techniques that are economical, eco-friendly, long-lasting, and rich in technology. Synthetic biology has indeed been utilized in the food, dairy, and agriculture areas, and has proven to have been a productive as well as efficient technology (Guazzaroni et al., 2015). This technology has enabled the development of foods with greater and superior nutritional characteristics, including those created for each individual need of the consumer, such as allergies or intolerances to certain foods. It has also facilitated the production of food with alterations to color, flavor, aroma, vitamin content, fiber, protein, fat, and carbohydrate content.

In addition, dietary approaches with high caloric value and fat, among several other aspects, have been designed to reduce them (Heffernan and Misturelli, 2000). Similarly, synthetic biology is being used to develop food packaging and biofilms to extend shelf life and retain food qualities (flavor, color, smell, and nutrients), as well as preserve them free of microbes that can harm them or cause pollution (Jensen and Keasling, 2015). On the other hand, because of the use of enzymes and compounds manufactured specifically for agriculture, this technique is also used to make fertilizers, plant development therapies, and pesticides. Using various molecular approaches to cure diseases or produce biosensors for the detection of pathogens, the food industry's production processes and food safety systems in the cultivation fields are also favored (Anderson et al., 2018; Chappell et al., 2015).

Furthermore, synthetic biology has been utilized to treat potent toxic pollutants like heavy metals, actinides, and nerve agents by making use of natural biodegradative mechanisms in particular bacteria. The use of bacteriophages created by bioengineering is another strategy for monitoring and removing harmful bacteria in cultures (Anderson et al., 2018). Because they precisely target and eliminate microbes of relevance, these viruses are non-harmful to humans, animals, and plants. Because pathogenic pages follow a lytic cycle in which they grow bacteria before inducing cell lysis, these mechanisms generate antibacterial action (Gutiérrez et al., 2016).

## 4.4 NANOTECHNOLOGY IN FOOD PROCESSING

Food produced employing nanotechnology in the processing, production, storage, and storage of food is referred to as nanofood. In post-harvest food production, nanotechnology provides enormous potential. It improves food bioavailability, flavor, texture, and consistency, or eliminates an unfavorable taste or odor, as well as changing particle size, size distribution, cluster formation potential, and surface charge (Powers et al., 2006). Edible nano-coatings (thin coatings of 5 nm) can be employed as gas and moisture barriers in meat, fruits, vegetables, cheese, fast food, bakery items, and confectionery products. They also give the processed items flavor, color, enzymes, antioxidants, anti-browning ingredients, and longer shelf life.

There are a number of baked foods with antibacterial nano-coatings that are edible (Azeredo et al., 2009; Naoto and Hiroshi, 2009). Nanofilters are being used to reduce color from beet juice while preserving the aroma and red wine, as well as to remove lactose from milk so that it can be replaced with other sugars, enabling the milk suitable for lactose-intolerant individuals. Nanoscale filters aid in the non-boiling elimination of bacterial species from milk or water. Nanomaterials used to make nanosieves are also used to filter milk and beer (Sekhon, 2014; Naoto and Hiroshi, 2009).

To prevent food diseases, nanotechnology is used in the production of nutritious foods that are low in cholesterol, sugar, or salt. Silicon dioxide ($SiO_2$) and titanium dioxide ($TiO_2$) were considered to be permitted as bulk food additives (E551 and E171, respectively) (EFSA, 2009). Bionanoencapsulated quercetin (biodegradable poly-D, L-lactide) has improved the storability of tomatoes, and this strategy is being used to preserve the quality of other fruit and vegetables (Yadav, 2017). The most prevalent commercialized nanotechnology-based products available are Nanogreen tea, Neosino capsules (dietary supplements), Canola active oil, Aquanova (micelle to enhance the solubility of vitamins (A, C, D, E, and K), beta-carotene, and omega fatty acids), and Nutralease (fortifying nanocarriers to carry nutraceuticals and drugs). Similarly, fortified fruit juices, oat nutritious beverages, nanoteas, nanocapsules containing tuna fish oil in bread, and nanoceuticals slim shakes are just a few examples of widely viable nano-processed foods that are commonly available in the United States, Australia, China, and Japan (Poças et al., 2008). Table 4.2 shows the use of nanotechnology in the production of commercialized nano food products as well as its applicability in other nutritional sciences techniques.

## 4.5 CONCLUSION

By minimizing inefficiency or limiting spoiling, food processing plays a vital role in addressing the issues of food availability. This overview emphasized the importance, usefulness, as well as present advancements in the plant, animal, and microbial bioengineered technologies, as well as viable strategies for enhancing the economic viability of food industries at the biological and process bioengineering levels. The

Table 4.2 Applications for Commercial Nanofood Products

| Product appellation | Category of product | Nanomaterial | Applications | References |
|---|---|---|---|---|
| Nutra Leaseanola Active Oi | Food and beverage | Nanosized self-assembled liquid structures (NSSL) | Inhibits transportation of cholesterol from the digestive system into the bloodstream | He et al. (2019) |
| Nanotea | Beverage | Nanoselenium | Good supplement of selenium | Nile et al. (2020) |
| Fortified Fruit Juice | Health drink | Micelles 5–100 nm in diameter | Increased Lycopene | |
| Nanoceuticals Slim Shake | Health drink | Conversion of vanilla or chocolate into nanoscale | Low-calorie diet | Pradhan et al. (2015) |
| NanoSlim beverage | Food and beverage | Liquid suspended nanoparticle | | |
| Nano B-12 Vitamin Spray | Food supplements | Nanodroplets | Efficacy enhancement | Pathakoti et al. (2017) |
| Neosino | Health supplement | Silicon | Health and fitness | Pathakoti et al. (2017) |
| Oat Chocolate and Oat Vanilla Nutritional Drink | Beverage | 300 nm of iron particles | Enhance bioavailability and reactivity | Walia et al. (2019) |
| Aquasol preservative | Food additive | Nanoscale micelle | Improves the absorption and effectiveness of dietary supplements and preservatives. | Rashidi and Khosravi-Darani (2011) |
| Omega-3 | Food additive | Nanocochleates as small as 50 nm | Efective addition of omega-3 fatty acids | Pawar (2016) |
| LycoVit | Food additive | < 200 nm synthetic lycopene | Effective antioxidant and used in soft drinks | Chatterjee (2016) |
| Antibacterial kitchenware | Food contact material | Nanoparticles of silve | Improved antibacterial properties | |
| Nano Silver Food Containers | Food storage | Silver | Storage | Nile et al. (2020) |
| Fresher Longer TM | Food storage | Plastic | Food products longevity | Nile et al. (2020) |
| Nanosilver cutting board | Food contact material | silver | Effective antibacterial | Chatterjee (2016) |

outcomes of the thorough scrutiny indicated that bioengineering techniques and approaches could have a positive effect on the food industry's development. Strategies for conquering and addressing the difficulties that come during the manufacture of food. It provides new insight into the problems and potential breakthroughs in the development of food bioproducts and new ingredients using bioengineered tools.

## REFERENCES

Aguilar CN, Ruiz HA, Rubio Rios A, Chávez- González M, Sepúlveda L, Rodríguez- Jasso RM, Loredo-Treviño A, Flores- Gallegos AC, Govea- Salas M, and Ascacio- Valdes JA. 2019. Emerging strategies for the development of food industries. *Bioengineered* 10(1): 522–537.

Ahmad V, Ahmad K, Baig MH, AL-Shwaiman HA, Al Khulaifi MM, Elgorban AM, and Khan MS. 2019. Efficacy of a novel bacteriocin isolated from *Lysinibacillus* sp. against Bacillus pumilus. *LWT* 102: 260–267.

Albano KM, and Nicoletti VR. 2018. Ultrasound impact on whey protein concentrate-pectin complexes and in the O/W emulsions with low oil soybean content stabilization. *Ultrason sonochem* 41: 562–571.

Anderson LA, Islam MA, and Prather KL. 2018. Synthetic biology strategies for improving the microbial synthesis of "green" biopolymers. *J Biol Chem* 293(14): 5053–5061.

Avaiyarasi ND, Ravindran AD, Venkatesh P, and Arul V. 2016. In vitro selection, characterization and cytotoxic effect of bacteriocin of Lactobacillus sakei GM3 isolated from goat milk. *Food Control* 69:1 24–33.

Azam SR, Ma H, Xu B, Devi S, Siddique MA, Stanley SL, Bhandari B, and Zhu J. 2020. Efficacy of ultrasound treatment in the removal of pesticide residues from fresh vegetables: A review. *Trends Food Sci Technol* 97: 417–432.

Azeredo HM, Mattoso LH, Wood D, Williams TG, Avena- Bustillos RJ, and McHugh TH. 2009. Nanocomposite edible films from mango puree reinforced with cellulose nanofibers. *J Food Sci* 74(5): N31–5.

Barba FJ. 2017. Microalgae and seaweeds for food applications: Challenges and perspectives. *Food Res Int* 99(3): 969–970.

Bhargav N, Mor RS, Kumar K, and Sharanagat VS. 2021. Advances in application of ultrasound in food processing: A review. *Ultrason sonochem* 70: 105293.

Briones-Labarca V, Giovagnoli-Vicuña C, and Chacana-Ojeda M. 2019. High pressure extraction increases the antioxidant potential and in vitro bio-accessibility of bioactive compounds from discarded blueberries. *CyTA-J Food* 17(1): 622–631.

Caraveo O, Alarcon- Rojo AD, Renteria A, Santellano E, and Paniwnyk L. 2015. Physicochemical and microbiological characteristics of beef treated with high-intensity ultrasound and stored at 4°C. *J Sci Food Agric* 95(12): 2487–2493.

Carcel JA, Castillo D, Simal S, and Mulet A. 2019. Influence of temperature and ultrasound on drying kinetics and antioxidant properties of red pepper. *Drying Technol* 37(4): 486–493.

Chappell J, Watters KE, Takahashi MK, and Lucks JB. 2015. A renaissance in RNA synthetic biology: new mechanisms, applications and tools for the future. *Curr Opin Chem Biol* 28: 47–56.

Chatterjee B. 2016. Synthetic Lycopene: the future but unaware fact. *Int J Clin Biomed Res* 2: 14–18.

Chen D, Cheng Y, Peng P, Liu J, Wang Y, Ma Y, Anderson E, Chen C, Chen P, Ruan R. Effects of intense pulsed light on *Cronobacter sakazakii* and *Salmonella surrogate Enterococcus faecium* inoculated in different powdered foods. *Food Chem* 296: 23–28.

Chen F, Zhang M, and Yang CH. 2020. Application of ultrasound technology in processing of ready-to-eat fresh food: A review. *Ultrason sonochem* 63: 104953.

Condón-Abanto S, Arroyo C, Álvarez I, Condón S, and Lyng JG. 2016. Application of ultrasound in combination with heat and pressure for the inactivation of spore forming bacteria isolated from edible crab (Cancer pagurus). *Int J Food Microbiol* 223: 9–16.

Cruz L, Clemente G, Mulet A, Ahmad-Qasem MH, Barrajón-Catalán E, and García-Pérez JV. 2016. Air-borne ultrasonic application in the drying of grape skin: Kinetic and quality considerations. *J Food Eng* 168: 251–258.

EFSA Scientific Committee. 2009. The potential risks arising from nanoscience and nanotechnologies on food and feed safety. *Sci Opin Sci Comm* 958: 1–39.

García P, Rodríguez L, Rodríguez A, and Martínez B. 2010. Food biopreservation: promising strategies using bacteriocins, bacteriophages and endolysins. *Trends Food Sci Technol* 21(8): 373–382.

Gayán E, Govers SK, and Aertsen A. 2017. Impact of high hydrostatic pressure on bacterial proteostasis. *Biophys Chem* 231: 3–9.

Guazzaroni ME, Silva-Rocha R, and Ward RJ. 2015. Synthetic biology approaches to improve biocatalyst identification in metagenomic library screening. *Micro Biotechnol* 8(1): 52–64.

Guimarães JT, Balthazar CF, Scudino H, Pimentel TC, Esmerino EA, Ashokkumar M, Freitas MQ, and Cruz AG. 2019. High-intensity ultrasound: A novel technology for the development of probiotic and prebiotic dairy products. *Ultrason Sonochem* 57: 12–21.

Gutiérrez D, Rodríguez-Rubio L, Martínez B, Rodríguez A, García P. 2016. Bacteriophages as weapons against bacterial biofilms in the food industry. *Front Microbiol* 7: 825.

He X, Deng H, and Hwang HM. 2019. The current application of nanotechnology in food and agriculture. *J Food Drug Anal* 27(1): 1–21.

Heffernan C, and Misturelli F. 2000. The delivery of veterinary services to the poor: Preliminary findings from Kenya. Report of the DFID Project. 7359.

Huang D, Men K, Li D, Wen T, Gong Z, Sunden B, and Wu Z. 2020. Application of ultrasound technology in the drying of food products. *Ultrason Sonochem* 63: 104950.

International Food Information Council, 2010.

Jensen MK, and Keasling JD. 2015. Recent applications of synthetic biology tools for yeast metabolic engineering. *FEMS Yeast Res* 15(1): 1–10.

Jo HL, Hwang HJ, and Chung MS. 2019. Inactivation of Bacillus subtilis spores at various germination and outgrowth stages using intense pulsed light. *Food Microbiol* 82: 409–415.

Ling B, Lyng JG, and Wang S. 2018. Radio-frequency treatment for stabilization of wheat germ: Dielectric properties and heating uniformity. *Innovative Food Sci Emerging Technol* 48: 66–74.

Martínez-López S, Lucas-Abellán C, Serrano-Martínez A, Mercader-Ros MT, Cuartero N, Navarro P, Pérez S, Gabaldón JA, and Gómez-López VM. 2019. Pulsed light for a cleaner dyeing industry: Azo dye degradation by an advanced oxidation process driven by pulsed light. *J Clean Prod* 217: 757–766.

Meyer RS, Cooper KL, Knorr D, and Lelieveld HL. 2000. High-pressure sterilization of foods. *Food Technol* 54(11): 67–129.

Misra NN, Schlüter O, and Cullen PJ. 2016. *Cold plasma in food and agriculture: fundamentals and applications*. Academic Press.

Morata Barrado A. 2010. *New Food Conservation Technologies*. Vol. 2. A. Madrid Vicente Ediciones.

Naoto S, and Hiroshi Mitsutoshi O. 2009. Micro-and nanotechnology for food processing. (Food safety series) resource: engineering and technology for sustainability. *World Am Soc Agric Eng* 16: 19.

Negi PS. 2012. Plant extracts for the control of bacterial growth: Efficacy, stability and safety issues for food application. *Int J Food Microbiol* 156(1): 7–17.

Nile SH, Baskar V, Selvaraj D, Nile A, Xiao J, and Kai G. 2020. Nanotechnologies in food science: applications, recent trends, and future perspectives. *Nano-micro lett* 12(1): 1–34.

Ozturk S, Kong F, Singh RK, Kuzy JD, Li C, and Trabelsi S. 2018. Dielectric properties, heating rate, and heating uniformity of various seasoning spices and their mixtures with radio frequency heating. *J Food Eng* 228: 128–141.

Ozturk S, Kong F, Singh RK, Kuzy JD, Li C. 2017. Radio frequency heating of corn flour: Heating rate and uniformity. *Innovative Food Sci Emerging Technol* 44: 191–201.

Pathakoti K, Manubolu M, and Hwang HM. 2017. Nanostructures: Current uses and future applications in food science. *J Food Drug Anal* 25(2): 245–253.

Pawar AY. 2016. Nanocochleate: a novel drug delivery system. Asian Journal of Pharmaceutics (AJP): Free full text articles from *Asian J Pharm* 10: 10(03).

Poças MF, Delgado TF, and Oliveira FA. 2008. Smart packaging technologies for fruits and vegetables. *Smart Packaging Technologies for Fast-Moving Consumer Goods*. 23: 151–166.

Powers KW, Brown SC, Krishna VB, Wasdo SC, Moudgil BM, and Roberts SM. 2006. Research strategies for safety evaluation of nanomaterials. Part VI. Characterization of nanoscale particles for toxicological evaluation. *Toxicol Sci* 90(2): 296–303.

Pradhan N, Singh S, Ojha N, Shrivastava A, Barla A, Rai V, and Bose S. 2015. Facets of nanotechnology as seen in food processing, packaging, and preservation industry. *BioMed Res Int* 2015: 1–17.

Quintero C, Falguera V, and Muñoz H. 2010. Films and edible coatings: importance, and recent trends in fruit and vegetable value chain. *Revista Tumbaga* 5(1): 93–118.

Rana A, and Meena S. 2017. Ultrasonic processing and its use in food industry: A review. *Int J Chem Stud* 5(6): 1961–1968.

Rashidi L, and Khosravi-Darani K. 2011. The applications of nanotechnology in the food industry. *Crit Rev Food Sci Nutr* 51(8): 723–730.

Salvucci E, Rossi M, Colombo A, Pérez G, Borneo R, and Aguirre A. 2019. Triticale flour films added with bacteriocin-like substance (BLIS) for active food packaging applications. *Food Pack Shelf Life* 19: 193–199.

Savadogo A, Ouattara AC, Bassole HI, and Traore SA. 2006. Bacteriocins and lactic acid bacteria-a minireview. *African J Biotechnol* 5(9): 678–683.

Sekhon BS. 2014. Nanotechnology in agri-food production: an overview. *Nanotechnol Sci Appl* 7: 31.

Settanni L, and Corsetti A. 2008. Application of bacteriocins in vegetable food biopreservation. *Int J Food Microbiol* 121(2): 123–138.

Walia N, Dasgupta N, Ranjan S, Ramalingam C, and Gandhi M. 2019. Food-grade nanoencapsulation of vitamins. *Environ Chem Lett* 17(2): 991–1002.

Wang A, Kang D, Zhang W, Zhang C, Zou Y, and Zhou G. 2018. Changes in calpain activity, protein degradation, and microstructure of beef M. semitendinosus by the application of ultrasound. *Food Chemistry* 245: 724–730.

Wang Y, Sun Y, Zhang X, Zhang Z, Song J, Gui M, Li P. 2015. Bacteriocin-producing probiotics enhance the safety and functionality of sturgeon sausage. *Food Control* 20: 729–735.

Xie F, Zhang W, Lan X, Gong S, Wu J, Wang Z. 2008. Effects of high hydrostatic pressure and high pressure homogenization processing on characteristics of potato peel waste pectin. *Carbohydr Polym* 196: 474–482.

Yadav SK. 2017. Tissue science & engineering realizing the potential of nanotechnology for agriculture and food technology. *J Tissue Sci Eng* 8: 8–11.

Zhang Y, and Abatzoglou N. 2020. Fundamentals, applications and potentials of ultrasound-assisted drying. *Chem Eng Res Des* 154: 21–46.

Zhu F, and Li H. (2019). Modification of quinoa flour functionality using ultrasound. *Ultrason Sonochem* 52: 305–310.

CHAPTER 5

# Functional Properties of Food Processing as a Novel Technology for Human Health and Nutrition

Ruby Angurana,[1#] Vaidehi Katoch,[2#] Tunisha Verma,[3] and Savita Bhardwaj[3*]
[1]Department of Zoology, School of Bioengineering and Biosciences,
Lovely Professional University, Phagwara (Punjab), India
[2]Department of Forensic Science, School of Bioengineering and Biosciences,
Lovely Professional University, Phagwara (Punjab), India
[3]Department of Botany, School of Bioengineering and Biosciences,
Lovely Professional University, Phagwara (Punjab), India
*Corresponding author email: savitakumaribhardwaj@gmail.com
# Equal Contribution

## CONTENTS

| | | |
|---|---|---|
| 5.1 | Introduction | 60 |
| 5.2 | Food Processing | 60 |
| 5.3 | Influence of the Food Industry on Nutrition | 63 |
| 5.4 | Effect of Food Processing on Human Health | 64 |
| | 5.4.1 Epidemiology Health Risks | 64 |
| | 5.4.2 The Nova Classification | 65 |
| | 5.4.3 The Food Matrix Influence on Health | 66 |
| 5.5 | Food Components and the Impact of Emerging Technologies | 66 |
| | 5.5.1 Proteins | 66 |
| | 5.5.2 Carbohydrates | 67 |
| | 5.5.3 Lipids | 68 |
| | 5.5.4 Essential Minerals | 68 |
| | 5.5.5 Vitamins | 69 |
| | 5.5.6 Enzymes | 69 |
| 5.6 | Future Aspects in Food Technology | 70 |
| 5.7 | Conclusion | 71 |
| References | | 72 |

DOI: 10.1201/9781003258568-5

## 5.1 INTRODUCTION

Food processing is ubiquitous and nearly every food item consumed by people is processed. Food processing exhibits both beneficial and harmful effects on human health. Modern systems of food are now changing across the world and currently the traditional food system is displaced by the ready to-consume food products (Moubarac et al., 2014). Humans have become acclimatized to various food systems through various methods such as food preparation and cooking, and food cultivation and preservation methods (Wrangham, 2013; Hotz and Gibson, 2007). Nutritional quality of food is remarkably improved by food preparation and cooking to increase the health benefits of food such as improved texture and taste, improved sensory characteristics and functional properties and an increase in the digestibility and accessibility of nutrients, vitamins, dietary fibers etc (Van Boekel et al., 2010). However, processed food is sometimes harmful as it exerts a deleterious impact on human health due to the formation of toxic compounds or by loss of nutrients (Seal et al., 2008; Hoffman and Gerber, 2015). Both harmful and advantageous effects are exerted by food processing on the food quality, depending as food processing results in alterations in the food components (Weaver et al., 2014).

Food processing connects agronomic production with the accessibility of food to the people from prehistoric times and is recognized as a chief feature of the food production chain (Floros et al., 2010). Alterations in the ingredients or food products in several ways are the basic techniques for food processing and to confirm the wellbeing, worth, and accessibility of consumable foods is the major function of food processing. Food processing is also utilized to improve the shelf-life of food products, to reduce food wastage and to retain nutrients, all resulting in sustainable food production system (MacDonald and Reitmeier, 2017). Processes such as milling, fermentation, drying, smoking, heating, canning, cooling/freezing, extrusion cooking etc. are some of the common food processing methods which are used in a commercial food system.

Non nutrient moieties, along with the food nutrition, are important parts of the food system in order to reduce the occurrence of diseases in humans (Sharma et al., 2016). A number of by-products are expected to be developed in the near future due to the quick increase in food processing trades, particularly in developing countries (Sharma et al., 2010; Joshi et al., 2011) and various bioactive compounds are present in high amounts in these by-products (Mckee et al., 2000). At both commercial and household level, use of thermal processes is required for the regulation of raw food products in the context of food safety and palatability issues. Sensory, nutritional and textural changes occur due to biological, physical and chemical modifications in food stuffs which are induced by the thermal treatment, which ultimately results in enhanced food safety and quality (Van Boekel et al., 2010).

## 5.2 FOOD PROCESSING

Food processing plays a vital role in feeding the growing global population. It is a series of unit activities that are used to transform raw food into foodstuffs with a longer shelf life and storage capability (Haouet et al., 2019), which eliminates or reduces the time and effort spent in culinary procedures, allowing for enhanced consumption of the food processed (Fróna et al., 2019). The availability of food has a considerable impact on human nutrition (Tripathi et al., 2019). This link translates into a considerable amount of influence for a broad and intricate company involved in the production and distribution of food in many contemporary communities in industrialized countries (El Bilali and Allahyari, 2018). Throughout history, aspects such as politics, economics, technology, the environment, and social issues have influenced activities in the food business sectors (Vågsholm et al., 2020). These factors have had an impact on efficiency, pricing, diversity, quality, and availability of items. When the food supply is numerous, affordable, and diverse, it is the decisions that people make that determine the nutritional quality of their meals, rather than the food supply itself (Vermeulen et al., 2020). As a result, the link between the food sector and public nutrition has a number of diverse aspects (Mackenbach et al., 2019).

The increased prevalence of being overweight and obesity, as well as the related health repercussions, has generated a global demand for altering eating habits (Popkin et al., 2020). Uncertainty exists regarding what the implications will be for the food business. The scope of a conversation considering these repercussions has the potential to be extraordinarily extensive. Choosing to concentrate on the implications of developments in nutrition science, paired with some significant social, environmental, and technological changes, throws some constraints on the spectrum of elements that can be addressed in this study. It is feasible to reflect on the influence of these aspects and provide some insights into the role of the food industry in promoting public nutrition into the twenty-first century by comparing past policy and practises of food producers with present initiatives (Ridgway et al., 2019). Most developing countries use the notion of food processing to improve the nutritional content of foods while also modifying its taste, scent, and texture in order to extend their shelf life and aesthetic features (Khan et al., 2018). In addition to being very perishable, high-quality items in high demand are also extremely perishable. With the help of modern technology, most perishable foods may be preserved, which is a blessing for those who live in a cold climate (Mercier et al., 2019).

With the successful implementation of commercial food preservation technology, the availability of perishable foods can be extended, thereby contributing to the general welfare of the human population (Ghoshal, 2018). While the demand for new processed items continues to grow, the fundamental principles of food processing remain the same in order to ensure long-term availability in times of scarcity. It is a constant challenge for the processed food industry to meet consumer expectations in order to create nutritious, pleasant, convenient and safe food items that are readily available and affordable (Sharif et al., 2018). When it comes to the production and distribution of food, food processing and value addition are key stages. In addition to making a significant contribution to the food value chain, developing food processing technologies that are both environmentally friendly and energy efficient can also aid in relieving the worldwide energy deficit (Martin-Rios et al., 2020). In spite of the fact that various food processing methods have been called into question in recent years, food processing cannot be avoided completely due to the ever-increasing human population that requires food. Food processing has been one of the most contentious issues in the food value chain, as a result of the world's increasing population over the previous two centuries (Qian et al., 2020).

According to estimates, as the world's population continues to grow, the demand for processed foods is expected to increase even further (Fróna et al., 2019). Due to numerous references from international organizations (e.g., the Food and Agriculture Organization of the United Nations and the World Health Organization) about chronic diseases caused by unhealthy lifestyle habits and unhealthy diets (Rampelli et al., 2018), processed foods are of interest to both the general public and researchers. There are many problems associated with lack of exercise and inactivity, including such things as the so-called "metabolic syndrome" and obesity (Jepsen et al., 2020). Both of these problems are important concerns in modern culture. Perhaps, the eating habits of the past two decades, which have been characterized by an increase in fast food consumption and highly processed meals, have had an impact on health (Janssen et al., 2017). The general public is becoming increasingly aware that proper nutrition is directly tied to their physical well-being and that it can help them prevent nutrition-related diseases (Witkamp, 2021). COVID-19 pandemic's severe implications have made immunity a key concern for modern consumers, who are keen to adapt their diets to include more healthy options in order to shield themselves from the pandemic's devastation (Bambra et al., 2020). A recent European survey of FMCC experts found that the vast majority of customers (up to 72 percent, according to the results of the survey) are looking for nutritious meals that not only have a balanced calorific content but also perform activities that are helpful to their overall health (Galanakis, 2021).

Agribusinesses have been using food processing to make edible, healthy and nutritious foods and to preserve raw agricultural products (Kothary and Mali, 2021). New non-thermal processing technologies produce foods that are microbiologically and chemically safe, while also enhancing the nutritional value, physical characteristics, and sensory attributes (Zhao et al., 2018). The implications of food processing at the microstructural level have only recently begun to be revealed, thanks to the development of microscopy tools and materials science concepts (Aguilera, 2018). This has led to the conclusion that processing (including cooking) is a carefully orchestrated effort to preserve, destroy, transform, and create edible structures, rather than an accident (Kabak, 2009). This technique resulted in the discovery of structure-property connections that were extended to texture, flavor, shelf life, product design, and nutritional content of food products (Roobab et al., 2021).

Nano-based food materials, innovative food packaging, intelligent delivery mechanisms of nutrients and bioactive materials, the implementation of green nanotechnologies for crop production, and the development and introduction of nano-biosensors to provide safer foods and waste reduction are all being developed and introduced (Pandhi et al., 2021). In the food industry, opportunities to exploit and develop nanotechnologies have resulted in a significant number of patents, as food technologists and engineers continue to explore creative ways to re-invent food products that will appeal to customers on a worldwide scale (Shafiq et al., 2020). When it comes to developing patentable technology, worries about consumer health and safety when nanoparticles are used in foods continues to be a source of contention (Karim, 2021).

These are the objectives of this chapter: to discuss the necessity and perspectives of food processing and preservation, history of food preservation, food spoilage, conventional and modern methods of food processing and preservation, characterization, and evaluation, industrialization to address food safety issues, food waste management, food security, and response to changing consumer demand, among other topics.

## 5.3 INFLUENCE OF THE FOOD INDUSTRY ON NUTRITION

Since the time of the early explorers, who founded small farming villages and established subsistence agriculture, a diverse range of storage and processing techniques has been adopted (Baiphethi and Jacobs, 2009). In addition to utilizing heat to cook or dry out the food, these methods require the addition of salt and sugar to the dish. Throughout the twentieth century, domestic and industrial refrigeration systems were created to preserve food by cooling and freezing it to a safe temperature (Xia and Sun, 2002). It was the later further development and integration of these technologies, which were also integrated with effective chemical preservatives, that resulted in enhanced manufacturing efficiencies and, as a result, increased food safety (Knorr et al., 2020). With the application of food technology to extend the shelf life of things, improve their appeal, utility, or novelty, customers became less educated about the procedures by which their food was produced and where it originated from, resulting in a decrease in consumer knowledge (Gutierrez et al., 2017).

Technology utilized in food processing has an impact on both its effectiveness and its amount of bio actives; for example, the more foods that are processed, stored, and transported, the greater the decline in the functional characteristics of the bio actives contained in such foods (Rawson et al., 2011). With the use of evolving technologies and other concerns in the food business (e.g., the urgent need for innovation and sustainability, food waste recovery, and so on), fresh data has been generated and the state of the art in the objectives linked to bioavailability has been advanced (Morone et al., 2019). There has been a dramatic shift in the way bio actives are incorporated into foods and consumed, which has had a big influence (Bharat Helkar and Sahoo, 2016). A food's nutritional value is typically connected with nutrients such as protein, lipids, and carbohydrates, as well as minor components (salt, a few vitamins such as iron, calcium and sodium, as well as additives and preservatives) that show on nutrition labels. It is less well known that in a product, these nutrients are neither homogeneously dispersed nor in a free form, but rather are found as part of complex microstructures that are difficult to detect (Moubarac et al., 2017).

As nutrition science has developed in recent decades, it has grown increasingly concerned not just with the types and amounts of nutrients required for good health, but also with the fraction of a particular vitamin that is really available for utilization by human bodies (Mozaffarian et al., 2018). A food's structural composition influences both the bio accessibility of nutrients (the fraction released during digestion) and the bioavailability of nutrients (the fraction that is actually absorbed) (Parada and Aguilera, 2007). In the small intestine, researchers observed several interactions with other food components, as well as the likelihood that the gut microbiota converts nutrients and bio actives released from the food matrix into beneficial metabolites before they are absorbed by the body (Palafox-Carlos et al., 2011). The major purpose of food processing is to lengthen the shelf life of products while also adding value to diets by providing people with food that is safe, convenient, diverse, and high in nutritional content (Jafarzadeh et al., 2021). It has been centuries since various unit operations and processes involving heat, mass and momentum transfer have been applied to a wide range of materials in order to achieve these goals, resulting in changes in the physical and chemical properties of foods as well as changes in biochemical and microbiological properties, alongside changes in organoleptic and nutritional properties of foods (Kotsanopoulos and Arvanitoyannis, 2013). A number of beneficial impacts of food processing have been reported, including an enhancement in flavor and texture as well as microbiological safety, in addition to an increase in digestibility and the bioavailability of specific minerals. Nutrition loss, protein aggregation growth, polymerization of oxidized lipids, and the production of several toxic compounds are all consequences of overheating (Samtiya et al., 2021).

The recommendations of nutritionists for the ingestion of the required quantities of nutrients through foods or supplements have made major contributions to the alleviation of several nutrient deficits during the course of the past century (Kalache et al., 2019). For example, scurvy and ascorbic acid are associated with pellagra and niacin, rickets and vitamin D are associated with thiamin, and neural tube abnormalities and folic acid are associated with niacin (Tulchinsky, 2010). However, if new technologies are utilized for an extended period of time, at a high intensity, or at particularly high temperatures, they may result in the breakdown of polymers (for example, proteins and carbohydrates) and the oxidation of labile substances (for example, lipids or glucosinolates) (Galanakis, 2021). Therefore, it is vital to regularly monitor and optimize the operational parameters of each individual component as a result of these considerations. The fact that many of these technologies have now been effectively demonstrated in an industrial context does not negate the need for additional research and development in this field (Freeman, 1996). Advances in modern procedures and food processing trends have led to significant advancements in the incorporation of ingredients, the fortification of foods and, ultimately, the manner in which we consume food. Accordingly, more in-depth studies are being conducted into the effects of developing technology on meals, with the goal of investigating various parameters (Weiss et al., 2010). In this spirit (e.g., bio accessibility characteristics and bioactivity of food compounds, nutritional value, applications, and shelf-life of foods, as well as their sensory aspects and consumer acceptance) (de Albuquerque et al., 2021).

## 5.4 EFFECT OF FOOD PROCESSING ON HUMAN HEALTH

### 5.4.1 Epidemiology Health Risks

The degree to which food has been processed is rarely taken into consideration in epidemiological studies. In epidemiological studies, foods are frequently categorized into groups such as fruits and vegetables, grains and legumes, dairy products and nuts, poultry, red and white meats, fish and shellfish, and so on (Fardet, 2018). The only comparisons that included processing were "red/processed meats" (which were subjected to both thermal and mechanical treatments), "whole/skimmed milks" (which were subjected to a mechanical treatment to remove the fat fraction), and "refined/whole grains" (which were subjected to a mechanical treatment to remove the bran and germ fraction (Holmboe-Ottesen and Wandel, 2012). Despite their limitations, these inexact comparisons give a preliminary peek into the role of processing in the development of chronic disease and mortality risks (Rosenthal and DiMatteo, 2001). A higher intake of whole grains has been linked to a lower risk of all-cause mortality, cardiovascular disease, cancer mortality, colorectal cancer, and lower fasting blood glucose and insulin levels, as well to a lesser extent with reduced weight gain and metabolic syndrome risk, whereas higher intake of refined grains has been linked to a higher risk of all-cause mortality, cardiovascular disease, cancer mortality, colorectal cancer, and lower fasting blood glucose and insulin levels, as well to a lesser extent with reduced weight (Rosenthal and DiMatteo, 2001).

According to research, refining cereal-based meals reduces their nutritional content. However, the researchers in these studies did not distinguish between naturally whole grain diets and whole grain foods manufactured from bran, germ, and/or cereal fiber (Huang et al., 2015). If ultra processed meals including fat and/or sugar have been blended with recombined from white refined flour, bran, and germ, they may be labelled as whole grain foods. There are various epidemiological studies that have been combined in meta-analyses for cancer, cardiovascular disease, obesity, and type 2 diabetes risks linked with red meats and processed meats (Jacobs et al., 2000). The metabolic syndrome is less well understood. In the latest analysis, red and processed meat was shown to be either neutral or related with an elevated risk of certain chronic conditions. The International Agency for Research on Cancer (IARC) has classified processed beef as a human carcinogen. One meta-analysis, however, discovered no association between red and processed meats and prostate cancer. Meta-analyses on the effects of red (non-processed) and processed meat (including ultra processed meat) found no discernible differences. Processed meat includes anything from traditional delicatessen cuts to ultra processed meats with various chemicals and additives (Raisuddin and Misra, 1991).

### 5.4.2 The Nova Classification

Recently published epidemiological studies using the NOVA categorization system have split people's consumption of calories from ultra processed meals into quartiles and quintiles (Martínez Steele et al., 2020). According to the researchers, those who had the most processed meals, regardless of food type, were more likely to be obese, to have metabolic syndrome, and to have high levels of dyslipidemia in their bloodstream. In a recent epidemiological study that looked at the availability of ultra-processed meals in homes across many countries, researchers discovered a robust association between the availability of processed foods and an increase in obesity prevalence of minor percentage point (Rauber et al., 2019).

### 5.4.3 The Food Matrix Influence on Health

Generally, the glycemic index is an excellent predictor of food processing since more processed foods have a higher glycemic index and a lower satiety potential (Fardet et al., 2018). Indeed, highly processed meals tend to be unstructured, fractionated, and enriched with free glucose and sucrose, which makes glucose more accessible for absorption and raises blood glycemic response (Fardet and Rock, 2019). The apple may have been the first food to be studied for its impact on glucose metabolism. Satiety and glycemic index of whole apples, apple juice, or apple puree were evaluated in healthy human subjects to see which was most beneficial (Davidou et al., 2020). Unstructured apples had a larger insulinemic peak 30 minutes after a meal and a lower satiety score compared to more structured apples. To replicate these findings, researchers manipulated the particle size of cereal (fine flour, coarse wheat flour, cracked grains, and whole grains) and found that finer cereal-based meal structures resulted in higher glucose and lower satiety levels (Venn and Mann, 2004). As with the satiety potential of carrot nutrients, the fiber and structure of whole carrots were compared to those without, and as expected, meals with whole and blended carrots resulted in significantly higher satiety than those with nutrients from carrots, emphasizing carrot's matrix effect (Moorhead et al., 2006). Whole and blended carrots were found to have a higher satiety potential than carrot nutrients.

Studies on breads cooked under various circumstances have shown that certain kinds of French bread had lower insulin indices in healthy people and lower glycemic indices in type 2 diabetic patients than those of the other varieties. These findings, say the authors, may be attributed to changes in bread processing rather than fiber content, showing once more a matrix impact related to processing (Arscott and Tanumihardjo, 2010).

## 5.5 FOOD COMPONENTS AND THE IMPACT OF EMERGING TECHNOLOGIES

### 5.5.1 Proteins

Bioactives in foods are losing their useful properties as they are processed, stored, and transported more and more (Galanakis, 2021). Bioavailability research has been boosted by the use of emerging technologies and other food industry concerns (e.g., the urgent need for innovations and sustainability, food waste recovery, etc.). In response to this evolution, bioactives are now being incorporated into food and consumed in novel ways. Amino acids are the building elements of protein, and they are connected together by substituted amide bonds (Hill et al., 2015). There are 20 distinct amino acids that make up the building blocks of dietary proteins. Protein may be obtained from a broad variety of sources, including meat, poultry, and eggs, among other things (Liskamp et al., 2011). These chemicals are found in high concentrations in a variety of foods, including meat and dairy products, fish, eggs, and legumes, as well as grains and oilseeds (Lundberg and Lindström, 2020). It is believed that animal protein is digested more completely than plant protein, and that this is why animal protein is preferred over plant protein. These physical and chemical properties include the essential amino acid sequence and composition, the distribution of charges in secondary and tertiary structures, the hydrophobicity/hydrophilicity ratio in secondary and tertiary structures and molecular flexibility; hydrodynamic properties such as texturization, gelation, and thickening; and finally topographical and physicochemical properties such as dispersibility and emulsifiability.

Food processing methods that are still in their infancy have been shown to have an impact on the content and function of dietary proteins, according to a number of research investigations (Sidnell and Greenstreet, 2011). It has been shown that ohmic heating may significantly reduce the breakdown of proteins (actin and myosin) in seafood and surimi (which is derived from squid mantle muscle), resulting in the creation of more stiff and elastic gels. The enzyme metalloproteinase, which is responsible for the breakdown of myosin during ordinary heating, was inactivated by heating the paste extremely quickly (from 0 to 90 degrees Celsius in less than a minute). However, the introduction of HPP causes the unfolding of dietary protein structures to occur more quickly (the main structure is not altered) by dissociating the non-covalent, ionic, hydrophobic, and hydrogen bonds that are present. The basic structure of dietary protein is not altered. As a consequence, it may be used not only for microbial inactivation, but also for protein stabilization, for example, by boosting the rate of enzymatic proteolysis, resulting in hydrolysates with lower residual antigenicity, and by limiting the development of bacteria, among other applications (Zimmermann and Mieth, 1986).

## 5.5.2 Carbohydrates

Carbohydrates are a class of chemically defined chemicals that have a wide range of physiological and physical qualities as well as health advantages, all of which are dependent on their source (Nazia Auckloo and Wu, 2016). Among the most important structural characteristics of carbohydrates are their polymerization degree and linkage type. These include water-soluble monosaccharides (e.g., D-glucose, D-mannose, D-galactose, L-arabinose, L-xylose, and D-fructose) and disaccharides (e.g., lactose and sucrose), as well as carbohydrates derived from individuals who consume dietary fiber are those who consume polymeric carbohydrates with ten or more monomeric units that have not been digested by the body's own enzymes in the small intestine. The human body makes use of the carbohydrates included in D-glucose, and practically all of them are digested and absorbed into the body with little effort (Goodman, 2010). Upon ingestion of a carbohydrate-containing meal, the glycaemic response (a corresponding increase and subsequent fall in blood glucose level) happens as a result of the rate of digestion and absorption of the food, as well as insulin's activity to bring the blood glucose level back into normal range (Elleuch et al., 2011).

Incorporating developing technology into food processing may maximize carbohydrate modification and extraction, resulting in novel food items with different physicochemical and functional qualities (e.g., modified starch). Ultrasounds have also been utilized to change tapioca starch, with the results showing that the starch's swelling power is boosted when the amplitude of the ultrasound is raised or the sonication period is extended (Nain et al., 2021). Corn was treated with PEF (with intensities up to 50 kV/cm) in order to determine the influence on starch characteristics. They claimed that as the intensity of the electric field increased, the temperature of gelatinization and viscosity of the solution dropped. Electroporation of sliced beetroots with pulse length and treatment period at 10 degrees Celsius was explored in order to increase the sucrose production of sugar beet. According to the findings, the acquired juice had greater sucrose content, while the concentrated juice was clearer and less colored than the juice obtained by diffusion in the standard method of extraction. When it comes to the recovery of insulin and oligo-fructose from various sources, ultrasound, microwaves, rapid solvent extraction, PEF, and ohmic heating have all been used, resulting in enhanced yields, decreased processing times, and reduced solvent and energy consumption (Askari et al., 2016).

### 5.5.3 Lipids

Many small lipid molecules (e.g., cholesterol, phytosterols) have been proven to have health-promoting qualities (such as Omega-3 and Omega-6 fatty acids) that may impact the body's physiological activities positively (Beardsell et al., 2002). When it comes to the physical properties of fats, triacylglycerols in fats play a key role, determining both the structure and type of the ordered phases. Fats aren't the most vulnerable to oxidation in food, but lipids are, and this is a major factor in food deterioration during processing and storage (Wills et al., 2006). In addition, lipids may be oxidized and degraded when they are consumed, digested, and finally absorbed into the intestinal lumen throughout their entire transit. Many of these non-thermal processing techniques include complex oxidation processes, thus it is important to consider the influence on lipids when using these procedures. It's important to keep an eye on all of the processing elements, such as treatment length, ultrasonic amplitude, PEF intensity and strength, and cold plasma flow frequency, in order to ensure proper treatment of HPP, ultrasound, and cold plasma. A study was conducted to determine whether or not cheddar cheese whey contained any volatile chemicals that were associated with the process of oxidative lipid oxidation. It has been shown that cold plasma processing, when used to decontaminate nuts such as walnuts and peanuts, may increase the lipids peroxide value by 20 percent when using greater power and for a longer treatment period (Thirumdas et al., 2014).

### 5.5.4 Essential Minerals

Minerals such as calcium, iron, and zinc are essential for a variety of bodily activities, and when they are not consumed in appropriate quantities, shortages manifest themselves as both specific and generic symptoms (Tardy et al., 2020). Minerals must be integrated into the diet, but their amount varies greatly across foods and dietary habits, despite the fact that they are present in diverse chemical forms and quantities. Calcium, for example, is mostly obtained through dairy products and other dairy products (Hunaiti and Saleh, 1996). Nonetheless, if the bioavailability is insufficient, the quantity of nutrients obtained by ingesting the foods listed above may not always be sufficient to meet dietary needs. The eating of fortified meals has been shown to increase the absorption of vital minerals in those who are deficient in such minerals. During the process of fortification of foods, the bio accessibility of minerals should be assessed, taking into consideration a variety of factors such as dietary inhibitors and promoters, as well as the methodology used in the assessment (McNeil and Lerner, 2013). There is relatively little information available on the influence of non-thermal methods on the availability of critical minerals. Non-thermal technologies, like conventional techniques, do not directly impact minerals, but they do cause changes in the physical characteristics and structure of the macromolecule that is connected with the non-thermal technology (Patel et al., 2017).

### 5.5.5 Vitamins

Vitamins are a collection of chemical components present in food that the body needs in order to function effectively (Finglas, 1999). Vitamins are essential for the body to function properly. There has been an increase in the desire for vitamin-dense foods and drinks that are minimally processed and high in nutrients, particularly in developed countries. Based on their solubility, vitamins may be separated into two groups: soluble vitamins (vitamins C and B) and fat-soluble vitamins (vitamins D, A, K, and E). It is necessary to increase the stability and bioavailability of vitamins in food in order to maintain their metabolic and health-promoting qualities (Fernández-García et al., 2009). Non-thermal approaches are utilized to accomplish this. Regardless, the degree to which irradiation changes Vitamin C is dependent on the kind of food that is being treated with radiation. Garlic and onions, for example, have no influence on the quantity of Vitamin C. In irradiated potatoes, the content of vitamin C decreased dramatically during the first several weeks of preservation. After a lengthy period of storage, the vitamin C levels in irradiated samples were higher than those in untreated controls. If the irradiation doses are kept to a bare minimum, the organoleptic characteristics of vitamins are not negatively impacted. Irradiation, on the other hand, has no impact on the antioxidants, vitamin B, K, and D, as well as on the carotenes (Prior et al., 2005).

### 5.5.6 Enzymes

An enzyme is a naturally occurring chemical catalyst that consists of a lengthy chain of amino acids (proteins). They adhere to a particular substrate location and work to reduce the amount of energy needed to trigger any process that occurs in living organisms (Seebach et al., 2005). Animals, plants, and microbes are all potential suppliers of enzymes. The concentration of the substrate and other physical parameters, like as pH and temperature, influence their activity. A permanent alteration in the enzyme's protein function alters the enzyme's overall performance. Thermal treatment is commonly used to deactivate enzymes, although this might result in unpleasant changes in the flavor of food (Rodrigues et al., 2011). As a result, new technologies like ohmic and microwave heating are becoming more commonplace. Polyphenol oxidases (PPOs), enzymes that cause fruit and vegetables to brown, have been inactivated, for example, by treating tender coconut water with ohmic heating. The deactivation of apple juice PPOs by ohmic heating was also verified. Solvent ionization modifies the enzyme surface charge, causing it to become inactive (Rodrigues et al., 2013). Applying HPP may also be used to deactivate enzymes, such as polygalacturonase, by causing structural and functional alterations to enzymes owing to the destruction of hydrogen bonds (Graham et al., 2014)

## 5.6 FUTURE ASPECTS IN FOOD TECHNOLOGY

In the future, nanotechnology has the potential to extend the shelf life of foods, personalize flavors, and improve human health and wellbeing in the food industry, among other things (Chandrapala and Leong, 2014). Nanotechnology has more problems than advantages, especially in terms of its influence on human health and the environment, and a broad variety of businesses will be at risk as a result of its deployment, with the bulk of concerns centered on the possible threats to human health (Adeyeye, 2019). When it comes to selling food, foods developed with nanotechnology that directly benefit customers will be easier to market than foods made with nanotechnology that do not immediately benefit consumers. It is reasonable for the industry to anticipate that sentiments on nanotechnology meals will become more favorable if a nanotechnology product with a desired benefit becomes available on the market (Cushen et al., 2012). The impacts of nanoparticles on human health are influenced by a variety of factors, including particle size and mass, chemical composition, surface characteristics, and the manner in which the individual nanoparticles aggregate together.

Nano scale materials are anticipated to pose the greatest risk to human health because of the areas where they enter, accumulate, and move inside the body. This is because they are so small. The use of nanotechnology in the food industry might be advantageous in the future. Nanotechnology, which involves the study and manipulation of matter at the nanoscale scale, enables products' material properties to be matched to their intended purpose and their utility to be increased by improving their usefulness (Senturk, 2013). These advancements include encapsulated ingredients that protect bioactives (e.g., omega-3 fatty acids, vitamins) while increasing nutrient delivery, nanomaterials for controlled delivery of anti-microbials, smart sensors for improved food safety management, and nanocomposite materials for improving barrier properties of packaging materials, among other things. When delivery system improvements improve the ingredient/product performance in their intended application, they may help to reduce the environmental impact of food production by using nutrients and functional components/actives more effectively.

Improved packaging materials and smart sensors may be able to assist in reducing waste in the supply chain by enabling more rapid response and intervention. Rethinking the potential of insects as a human food source is necessary since they are currently consumed in many nations. Natural resource conservation measures are needed to ensure the long-term viability of wild collection of edible insects. It is possible to enhance insect populations by modifying their habitat (Ravichandran, 2010). Pest insects may be used as a source of food and feed at the same time, which is a win-win situation. Some potential insect species need the development of simple rearing techniques. Since iron and zinc deficits are so prevalent in the tropical regions, greater research is needed into the bioavailability of these nutrients in edible insects. Automated mass-rearing facilities that generate stable, consistent, and safe products are required to replace present protein-rich sources like meal and oil from fish and soybeans. As a new industry, this one has the difficulty of producing a high-quality, cost-effective insect biomass. It is necessary to create formal regulatory structures. Success will only be possible if the government, industry, and academics work together (Kunisho et al., 2010).

## 5.7 CONCLUSION

Toxins from food contact goods, such as food packaging, migrate into the food supply and are ingested by humans. Many of these chemicals haven't been properly explored for their possible health effects, while others have been confirmed to be dangerous to humans. Consequently, we recognize the need to change how migratory compounds' safety is assessed in light of the most recent scientific findings. Meanwhile, a wide range of groups are attempting to find ways to reduce packaging waste and reduce plastic pollution without considering the safety of chemicals. All stakeholders should focus more on this issue and use science-based decision making in the interest of improving public health. As a preventative measure, reducing the amount of harmful food contact chemicals in people's diets is beneficial. For this reason, it is important to include chemical safety concerns in the design of environmentally friendly packaging. However, the application of emerging technologies may result in the degradation of polymers, for example, proteins and carbohydrates and the oxidation of labile compounds for example, lipids or glucosinolates if the technology is used for an extended period of time, at a high intensity, or at relatively high temperatures. As a result, it is necessary to continuously monitor and optimize operating parameters for each individual component. Despite the fact that many of these technologies have already been successfully proven in the industrial setting, further efforts are required in this regard. Modern methods and food processing trends have resulted in breakthroughs in the inclusion of components, the fortification of meals, and, ultimately, the manner in which we consume food.

## REFERENCES

Adeyeye SAO. 2019. Food packaging and nanotechnology: safeguarding consumer health and safety. *Nutr Food Sci* doi:10.1108/nfs-01-2019-0020

Aguilera JM. 2018. The food matrix: implications in processing, nutrition and health. *Crit Rev Food Sci Nutr* 59: 3612–3629. doi:10.1080/10408398.2018.1502743

Arscott SA, and Tanumihardjo SA. 2010. Carrots of many colors provide basic nutrition and bioavailable phytochemicals acting as a functional food. *Compr Rev Food Sci Food Saf* 9: 223–239. doi:10.1111/j.1541-4337.2009.00103.x

Askari G, Asgharian A, Esmailzade A, Feizi A, and Mohammadi V. 2016. The effect of symbiotic supplementation on liver enzymes, c-reactive protein and ultrasound findings in patients with non-alcoholic fatty liver disease: A clinical trial. *Int J Prev Med* 7: 59. doi:10.4103/2008-7802.178533

Baiphethi MN, and Jacobs PT. 2009. The contribution of subsistence farming to food security in South Africa. *Agrekon* 48: 459–482. doi:10.1080/03031853.2009.9523836

Bambra C, Riordan R, Ford J, and Matthews F. 2020. The COVID-19 pandemic and health inequalities. *J Epidemiol Commun Health.* doi:10.1136/jech-2020-214401

Beardsell D, Francis J, Ridley D, and Robards K. 2002. Health Promoting Constituents in Plant Derived Edible Oils. *J Food Lipids* 9: 1–34. doi:10.1111/j.1745-4522.2002.tb00205.x

Bharat Helkar P, and Sahoo A. 2016. Review: Food Industry By-Products used as a Functional Food Ingredients. *Int J Waste Resour* 6. doi:10.4172/2252-5211.1000248

Chandrapala, J, and Leong T. 2014. Ultrasonic Processing for Dairy Applications: Recent Advances. *Food Eng Rev* 7: 143–158. doi:10.1007/s12393-014-9105-8

Cushen M, Kerry J, Morris M, Cruz- Romero M, and Cummins E. 2012. Nanotechnologies in the food industry—Recent developments, risks and regulation. *Trends Food Sci Technol* 24: 30–46. doi:10.1016/j.tifs.2011.10.006

Davidou S, Christodoulou A, Fardet A, and Frank K. 2020. The holistico-reductionist Siga classification according to the degree of food processing: an evaluation of ultra-processed foods in French supermarkets. *Food Function* 11: 2026–2039. doi:10.1039/c9fo02271f

de Albuquerque JG, Escalona-Buendía HB, de Magalhães Cordeiro AMT, dos Santos Lima M, de Souza Aquino J, and da Silva Vasconcelos MA. 2021. Ultrasound treatment for improving the bioactive compounds and quality properties of a Brazilian nopal (Opuntia ficus-indica) beverage during shelf-life. *LWT* 149: 111814. doi:10.1016/j.lwt.2021.111814

El Bilali H, and Allahyari MS. 2018. Transition towards sustainability in agriculture and food systems: Role of information and communication technologies. *Inf Process Agric* 5: 456–464. doi:10.1016/j.inpa.2018.06.006

Elleuch M, Bedigian D, Roiseux, O, Besbes S, Blecker C, and Attia H. 2011. Dietary fibre and fibre-rich by-products of food processing: Characterisation, technological functionality and commercial applications: A review. *Food Chem* 124: 411–421. doi:10.1016/j.foodchem.2010.06.077

Fardet A, and Rock E. 2019. Ultra-processed foods: A new holistic paradigm? *Trends Food Sci Technol* 93: 174–184. doi:10.1016/j.tifs.2019.09.016

Fardet A, Lakhssassi S, and Briffaz A. 2018. Beyond nutrient-based food indices: a data mining approach to search for a quantitative holistic index reflecting the degree of food processing and including physicochemical properties. *Food Function* 9: 561–572. doi:10.1039/c7fo01423f

Fardet A. 2018. Characterization of the degree of food processing in relation with its health potential and effects. *Adv Food Nutr Res* 79–129. doi:10.1016/bs.afnr.2018.02.002

Fernández-García E, Carvajal-Lérida I, and Pérez-Gálvez A. 2009. In vitro bioaccessibility assessment as a prediction tool of nutritional efficiency. *Nutr Res* 29: 751–760. doi:10.1016/j.nutres.2009.09.016

Finglas P. 1999. Bioavailability and analysis of vitamins in foods. *Eur J Clin Nutr* 53: 80–81. doi:10.1038/sj.ejcn.1600596

Floros JD, Newsome R, Fisher W, Barbosa-Cánovas GV, Chen H, Dunne CP, German JB, Hall RL, Heldman DR, Karwe MV, and Ziegler GR. 2010. Feeding the world today and tomorrow: the importance of food science and technology: an IFT scientific review. *Compr Rev Food Sci and Food Saf* 9(5): 572–599.

Freeman C. 1996. The greening of technology and models of innovation. *Technol Forecast Soc Change* 53: 27–39. doi:10.1016/0040-1625(96)00060-1

Fróna D, Szenderák J, and Harangi-Rákos M. 2019. The challenge of feeding the world. *Sustainability* 11: 5816. doi:10.3390/su11205816

Galanakis CM. 2021. Functionality of food components and emerging technologies. *Foods* 10: 128. doi:10.3390/foods10010128

Ghoshal G. 2018. Emerging food processing technologies. *Food Processing for Increased Quality and Consumption* 29–65. doi:10.1016/b978-0-12-811447-6.00002-3

Goodman BE. 2010. Insights into digestion and absorption of major nutrients in humans. *Adv Physiol Educ* 34: 44–53. doi:10.1152/advan.00094.2009

Graham JD. Buytendyk AM, Wang D, Bowen KH, and Collins KD. 2014. Strong, low-barrier hydrogen bonds may be available to enzymes. *Biochemistry* 53: 344–349. doi:10.1021/bi4014566

Gutierrez MM, Meleddu M, and Piga, A. 2017. Food losses, shelf life extension and environmental impact of a packaged cheesecake: A life cycle assessment. *Food Res Int* 91: 124–132. doi:10.1016/j.foodres.2016.11.031

Haouet MN, Tommasino M, Mercuri ML, Benedetti F, Di Bella S, Framboas M, Pelli S, and Altissimi MS. 2019. Experimental accelerated shelf life determination of a ready-to-eat processed food. *Ital J Food Saf* 7. doi:10.4081/ijfs.2018.6919

Hill TA, Shepherd NE, Diness, F, and Fairlie DP. 2015. ChemInform abstract: constraining cyclic peptides to mimic protein structure motifs. *ChemInform* 46: no-no. doi:10.1002/chin.201508304

Hoffman R, and Gerber M. 2015. Food processing and the Mediterranean diet. *Nutrients* 7(9): 7925–7964.

Holmboe-Ottesen G, and Wandel M. 2012. Changes in dietary habits after migration and consequences for health: a focus on South Asians in Europe. *Food Nutr Res* 56: 18891. doi:10.3402/fnr.v56i0.18891

Hotz C, and Gibson RS. 2007. Traditional food-processing and preparation practices to enhance the bioavailability of micronutrients in plantbased diets. *J Nutr* 137(4): 1097–1100.

Huang T, Xu M, Lee A, Cho S, and Qi L. 2015. Erratum: Consumption of whole grains and cereal fiber and total and cause-specific mortality: prospective analysis of 367,442 individuals. *BMC Med* 13. doi:10.1186/s12916-015-0338-z

Hunaiti AA, and Saleh MS. 1996. Effects of iron, zinc, calcium, and vitamins on the activity and contents of human placental copper/zinc and manganese superoxide dismutases. *Biol Trace Elem Res* 54: 231–238. doi:10.1007/bf02784434

Jacobs DR, Pereira MA, Meyer KA, and Kushi LH. 2000. Fiber from whole grains, but not refined grains, is inversely associated with all-cause mortality in older women: The Iowa Women's Health Study. *J Am Coll Nutr* 19: 326S330S. doi:10.1080/07315724.2000.10718968

Jafarzadeh S, Mohammadi Nafchi A, Salehabadi A, Oladzad-abbasabadi N, and Jafari SM. 2021. Application of bio-nanocomposite films and edible coatings for extending the shelf life of fresh fruits and vegetables. *Adv Colloid Interface Sci* 291: 102405. doi:10.1016/j.cis.2021.102405

Janssen HG, Davies IG, Richardson LD, and Stevenson L. 2017. Determinants of takeaway and fast food consumption: a narrative review. *Nutr Res Rev* 31: 16–34. doi:10.1017/s0954422417000178

Jepsen S, Suvan J, and Deschner J. 2020. The association of periodontal diseases with metabolic syndrome and obesity. *Periodontology* 2000 83: 125–153. doi:10.1111/prd.12326

Joshi VK, and Sharma SK. 2011. Food processing industrial waste–present scenario. In: *Food Processing Waste Management: Treatment and Utilization Technology*, New India Pub. Agency, New Delhi, 1–30.

Kabaks B. 2009. The fate of mycotoxins during thermal food processing. *J Sci Food and Agric* 89: 549–554. doi:10.1002/jsfa.3491

Kalache A, de Hoogh AI, Howlett SE, Kennedy B, Eggersdorfer M, Marsman DS, Shao A, and Griffiths JC. 2019. Nutrition interventions for healthy ageing across the lifespan: a conference report. *Euro J Nutr* 58: 1–11. doi:10.1007/s00394-019-02027-z

Karim ME. 2021. Nanoproducts and legal aspects of consumer protections: an evaluation. *Handbook of Consumer Nanoproducts*: 1–26. doi:10.1007/978-981-15-6453-6_79-1

Khan MK, Ahmad K, Hassan S, Imran M, Ahmad N, and Xu C. 2018. Effect of novel technologies on polyphenols during food processing. *Innov Food Sci Emerg Technol* 45: 361–381. doi:10.1016/j.ifset.2017.12.006

Knorr D, Augustin MA, and Tiwari B. 2020. Advancing the Role of Food Processing for Improved Integration in Sustainable Food Chains. *Front Nutr* 7. doi:10.3389/fnut.2020.00034

Kothary S, and Mali A. 2021. Improving sustainability in Indian cities through expansion of edible green spaces: exploring million plus cities of Bengaluru, Hyderabad and Ahmedabad. *Towards Implementation of Sustainability Concepts in Developing Countries*: 113–128. doi:10.1007/978-3-030-74349-9_9

Kotsanopoulos KV, and Arvanitoyannis IS. 2013. Membrane processing technology in the food industry: food processing, wastewater treatment, and effects on physical, microbiological, organoleptic, and nutritional properties of foods. *Crit Rev Food Sci Nutr* 55: 1147–1175. doi:10.1080/10408398.2012.685992

Kunisho S, Noguchi T, and Takano K. 2010. Effects of some reagents on the intermolecular interactions among soybean proteins during the formation of yuba-films. *Food Preservation Science* 36: 131–133. doi:10.5891/jafps.36.131

Liskamp RMJ, Rijkers DTS, Kruijtzer JAW, and Kemmink J. 2011. Peptides and proteins as a continuing exciting source of inspiration for peptidomimetics. *Chem Bio Chem* 12: 1626–1653. doi:10.1002/cbic.201000717

Lundberg C, and Lindström KN. 2020. Sustainable management of popular culture tourism destinations: a critical evaluation of the twilight saga servicescapes. *Sustainability* 12: 5177. doi:10.3390/su12125177

MacDonald R, and Reitmeier C. 2017. *Food Processing. Understanding food systems: agriculture, food science, and nutrition in the United States*. Academic Press.

Mackenbach JD, Nelissen KGM, Dijkstra SC, Poelman MP, Daams JG, Leijssen JB, and Nicolaou M. 2019. A systematic review on socioeconomic differences in the association between the food environment and dietary behaviors. *Nutrients* 11: 2215. doi:10.3390/nu11092215

Martínez Steele E, Khandpur N, da Costa Louzada ML, and Monteiro CA. 2020. Association between dietary contribution of ultra-processed foods and urinary concentrations of phthalates and bisphenol in a nationally representative sample of the US population aged 6 years and older. *PLOS ONE* 15: e0236738. doi:10.1371/journal.pone.0236738

Martin- Rios C, Hofmann A, and Mackenzie N. 2020. Sustainability-oriented innovations in food waste management technology. *Sustainability* 13: 210. doi:10.3390/su13010210

Mckee L, and Latner TA. 2000. Underutilized sources of dietary fiber: A review. *Plant Foods Human Nutr* 55(4): 285–304.

McNeil J, and Lerner AB. 2013. Standard thromboelastography should not be used to assess candidacy for neuraxial procedures in patients taking P2Y12 inhibitors. *Anesthesiology* 119: 993. doi:10.1097/aln.0b013e3182a44648

Mercier S, Mondor M, McCarthy U, Villeneuve S, Alvarez G, and Uysal I. 2019. Optimized cold chain to save food. *Saving Food*: 203–226. doi:10.1016/b978-0-12-815357-4.00007-9

Moorhead SA, Welch RW, Barbara M, Livingstone E, McCourt M, Burns AA, and Dunne A. 2006. The effects of the fibre content and physical structure of carrots on satiety and subsequent intakes when eaten as part of a mixed meal. *Brit J Nutr* 96: 587–595. doi:10.1079/bjn20061790

Morone P, Koutinas A, Gathergood N, Arshadi M, and Matharu A. 2019. Food waste: Challenges and opportunities for enhancing the emerging bio-economy. *J Clean Prod* 221: 10–16. doi:10.1016/j.jclepro.2019.02.258

Moubarac JC, Batal M, Louzada ML, Martinez Steele E, and Monteiro CA. 2017. Consumption of ultra-processed foods predicts diet quality in Canada. *Appetite* 108: 512–520. doi:10.1016/j.appet.2016.11.006

Moubarac JC, Parra DC, Cannon G, and Monteiro CA. 2014. Food classification systems based on food processing: significance and implications for policies and actions: a systematic literature review and assessment. *Curr Obes Rep* 3(2): 256–272.

Mozaffarian D, Rosenberg I, and Uauy R. 2018. History of modern nutrition science—implications for current research, dietary guidelines, and food policy. *BMJ*: k2392. doi:10.1136/bmj.k2392

Nain N, Katoch, GK, Kaur S, and Rasane P. 2021. Recent developments in edible coatings for fresh fruits and vegetables. *J Hortic Res* 29: 127–140. doi:10.2478/johr-2021-0022

Nazia Auckloo B, and Wu B. 2016. Structure, biological properties and applications of marine-derived polysaccharides. *Curr Org Chem* 20: 2002–2012. doi:10.2174/1385272820666160202003944

Palafox-Carlos H, Ayala-Zavala JF, and González-Aguilar GA. 2011. The role of dietary fiber in the bioaccessibility and bioavailability of fruit and vegetable antioxidants. *J Food Sci* 76: R6–R15. doi:10.1111/j.1750-3841.2010.01957.x

Pandhi S, Mahato DK, and Kumar A. 2021. Overview of green nanofabrication technologies for food quality and safety applications. *Food Rev Int* 1–21. doi:10.1080/87559129.2021.1904254

Parada J, and Aguilera JM. 2007. Food microstructure affects the bioavailability of several nutrients. *J Food Sci* 72: R21–R32. doi:10.1111/j.1750-3841.2007.00274.x

Patel JJ, Mundi MS, Hurt RT, Wolfe B, and Martindale RG. 2017. Micronutrient deficiencies after bariatric surgery: an emphasis on vitamins and trace minerals. *Nutr Clin Pract* 32: 471–480. doi:10.1177/0884533617712226

Popkin BM, Corvala, C, and Grummer-Strawn LM. 2020. Dynamics of the double burden of malnutrition and the changing nutrition reality. *Lancet* 395: 65–74. doi:10.1016/s0140-6736(19)32497-3

Prior RL, Wu X, and Schaich K. 2005. Standardized methods for the determination of antioxidant capacity and phenolics in foods and dietary supplements. *J Agric Food Chem* 53: 4290–4302. doi:10.1021/jf0502698

Qian J, Dai B, Wang B, Zha Y, and Song Q. 2020. Traceability in food processing: problems, methods, and performance evaluations—a review. *Crit Rev Food Sci Nutr* 1–14. doi:10.1080/10408398.2020.1825925

Raisuddin S, and Misra JK. 1991. Aflatoxin in betel nut and its control by use of food preservatives. *Food Addit Contam* 8: 707–712. doi:10.1080/02652039109374028

Rampelli S, Guenther K, Turroni S, Wolters M, Veidebaum T, Kourides Y, Molnár D, Lissner L, Benitez-Paez A, Sanz Y, Fraterman A, Michels N, Brigidi P, Candela M, and Ahrens W. 2018. Pre-obese children's dysbiotic gut microbiome and unhealthy diets may predict the development of obesity. *Commun Biol* 1. doi:10.1038/s42003-018-0221-5

Rauber F, Louzada ML da C, Martinez Steele E, Rezende LFM de, Millett C, Monteiro CA, and Levy RB. 2019. Ultra-processed foods and excessive free sugar intake in the UK: a nationally representative cross-sectional study. *BMJ Open* 9: e027546. doi:10.1136/bmjopen-2018-027546

Ravichandran R. 2010. Nanotechnology applications in food and food processing: innovative green approaches, opportunities and uncertainties for global market. *Int J Green Nanotechnol Phy Chem* 1: P72–P96. doi:10.1080/19430871003684440

Rawson A, Patras A, Tiwari BK, Noci F, Koutchma T, and Brunton N. 2011. Effect of thermal and non thermal processing technologies on the bioactive content of exotic fruits and their products: Review of recent advances. *Food Res Int* 44: 1875–1887. doi:10.1016/j.foodres.2011.02.053

Ridgway E, Baker P, Woods J, and Lawrence M. 2019. Historical developments and paradigm shifts in public health nutrition science, guidance and policy actions: a narrative review. *Nutrients* 11: 531. doi:10.3390/nu11030531

Rodrigue RC, Berenguer- Murcia A, and Fernandez- Lafuente R. 2011. ChemInform abstract: coupling chemical modification and immobilization to improve the catalytic performance of enzymes. *ChemInform* 43: no-no. doi:10.1002/chin.201202266

Rodrigues RC., Ortiz C, Berenguer- Murcia A, Torres R, and Fernandez- Lafuente R. 2013. ChemInform abstract: modifying enzyme activity and selectivity by immobilization. *ChemInform* 44: no-no. doi:10.1002/chin.201339264

Roobab U, Inam- Ur- Raheem M, Khan AW, Arshad RN, Zeng X, and Aadil RM. 2021. Innovations in high-pressure technologies for the development of clean label dairy products: a review. *Food Rev Int* 1–22. doi:10.1080/87559129.2021.1928690

Rosenthal R, and DiMatteo MR. 2001. Meta-analysis: recent developments in quantitative methods for literature reviews. *Annu Rev Psychol* 52: 59–82. doi:10.1146/annurev.psych.52.1.59

Samtiya M, Aluko RE, Puniya AK, and Dhewa T. 2021. Enhancing micronutrients bioavailability through fermentation of plant-based foods: a concise review. *Fermentation* 7: 63. doi:10.3390/fermentation7020063

Seal CJ, de Mul A, Eisenbrand G, Haverkort AJ, Franke K, Lalljie SPD, Mykkänen H, Reimerdes E, Scholz G, Somoza V, and Wilms L. 2008. Risk-benefit considerations of mitigation measures on acrylamide content of foods–a case study on potatoes, cereals and coffee. *Brit J Nutr* 99: S1-S46.

Seebach, D, Beck AK, and Bierbaum DJ. 2005. The world of β- and γ-peptides comprised of homologated proteinogenic amino acids and other components. *ChemInform* 36. doi:10.1002/chin.200533334

Senturk A. 2013. Nanotechnology as a food perspective. *J Nanomater Mol Nanotechnol* 02. doi:10.4172/2324-8777.1000125

Shafiq M, Anjum S, Hano C, Anjum I, and Abbasi BH. 2020. An overview of the applications of nanomaterials and nanodevices in the food industry. *Foods* 9: 148. doi:10.3390/foods9020148

Sharif MK, Zahid A, and Shah FH. 2018. Role of food product development in increased food consumption and value addition. *Food Processing for Increased Quality and Consumption*. 455–479. doi:10.1016/b978-0-12-811447-6.00015-1

Sharma SK, Bansal S, Mangal M, Dixit AK, Gupta RK, and Mangal AK. 2016. Utilization of food processing by-products as dietary, functional, and novel fiber: a review. *Crit Rev Food Sci Nutr* 56(10): 1647–1661

Sharma SK. 2010. Functional foods and nutraceuticals. In: *Postharvest Management and Processing of Fruits and Vegetables—Instant Notes*, New India Pub. Agency, New Delhi, 283–288.

Sidnell A, and Greenstreet E. 2011. Infant nutrition—review of lipid innovation in infant formula. *Nutr Bull* 36: 373–380. doi:10.1111/j.1467-3010.2011.01913.x

Tardy A-L, Pouteau E, Marquez D, Yilmaz C, and Scholey A. 2020. Vitamins and minerals for energy, fatigue and cognition: a narrative review of the biochemical and clinical evidence. *Nutrients* 12: 228. doi:10.3390/nu12010228

Thirumdas R, Sarangapani C, and Annapure US. 2014. Cold plasma: a novel non-thermal technology for food processing. *Food Biophysics* 10: 1–11. doi:10.1007/s11483-014-9382-z

Tripathi AD, Mishra R, Maurya KK, Singh RB, and Wilson DW. 2019. Estimates for world population and global food availability for global health. *The Role of Functional Food Security in Global Health*. 3–24. doi:10.1016/b978-0-12-813148-0.00001-3

Tulchinsky TH. 2010. Micronutrient deficiency conditions: global health issues. *Public Health Rev* 32: 243–255. doi:10.1007/bf03391600

Vågsholm I, Arzoomand NS, and Boqvist S. 2020. Food security, safety, and sustainability—getting the trade-offs right. *Front Sustain Food Syst* 4. doi:10.3389/fsufs.2020.00016

Van Boekel M, Fogliano V, Pellegrini N, Stanton C, Scholz G, Lalljie S, Somoza V, Knorr D, Jasti PR, and Eisenbrand G. 2010. A review on the beneficial aspects of food processing. *Mol Nutr and Food Res* 54(9): 1215–1247.

Venn BJ, and Mann JI. 2004. Cereal grains, legumes and diabetes. *Euro J Clin Nutr* 58: 1443–1461. doi:10.1038/sj.ejcn.1601995

Vermeulen SJ, Park T, Khoury CK, and Béné C. 2020. Changing diets and the transformation of the global food system. *Ann N Y Acad Sci* 1478: 3–17. doi:10.1111/nyas.14446

Weaver CM, Dwyer J, Fulgoni III VL, King JC, Leveille GA, MacDonald RS, Ordovas J, and Schnakenberg D. 2014. Processed foods: contributions to nutrition. *Am J Clin Nutr* 99(6): 1525–1542.

Weiss J, Gibis M, Schuh V, and Salminen H. 2010. Advances in ingredient and processing systems for meat and meat products. *Meat Sci* 86: 196–213. doi:10.1016/j.meatsci.2010.05.008

Wills TM, Dewitt CAM, Sigfusson H, and Bellmer D. 2006. Effect of cooking method and ethanolic tocopherol on oxidative stability and quality of beef patties during refrigerated storage (oxidative stability of cooked patties). *J Food Sci* 71: C109–C114. doi:10.1111/j.1365-2621.2006.tb15604.x

Witkamp RF. 2021. Nutrition to optimise human health—how to obtain physiological substantiation? *Nutrients* 13: 2155. doi:10.3390/nu13072155

Wrangham R. 2013. The evolution of human nutrition. *Curr Biol* 23(9): 354–355.

Xia B, and Sun DW. 2002. Applications of computational fluid dynamics (cfd) in the food industry: a review. *Comput Electron Agric* 34: 5–24. doi:10.1016/s0168-1699(01)00177-6

Zhao YM, de Alba M, Sun DW, and Tiwari B. 2018. Principles and recent applications of novel non-thermal processing technologies for the fish industry—a review. *Crit Rev Food Sci Nutr* 59: 728–742. doi:10.1080/10408398.2018.1495613

Zimmermann R, and Mieth G. 1986. Selected modification of plant proteins and their functionality regarding wheat dough rheology. *Food Nahrung* 30: 439–441. doi:10.1002/food.19860300375

CHAPTER 6

# Food Processing Potential for Energy Efficiency and Use

**Dhriti Sharma,[1] Savita Bhardwaj,[1] Tunisha Verma,[1] Mamta Pujari,[1] Rahul Singh,[2] and Vandana Gautam[3*]**

[1]Department of Botany, School of Bioengineering and Biosciences,
Lovely Professional University, Phagwara (Punjab), India
[2]Department of Zoology, School of Bioengineering and Biosciences,
Lovely Professional University, Phagwara (Punjab), India
[3]College of Horticulture and Forestry, Dr. Y. S. Parmar University of Horticulture and Forestry, Nauni, Solan (HP), Neri Campus (Himachal Pradesh), India
*Corresponding author email: vndu.gndu@gmail.com

## CONTENTS

- 6.1 Introduction .................................................................................. 80
- 6.2 Tracing the Journey of Food Processing ...................................... 81
- 6.3 Food Processing Potential ............................................................ 82
- 6.4 Energy Consumption Patterns in Food Processing ...................... 83
- 6.5 Food Processing: Energy Consuming Techniques at a Glance .... 83
- 6.6 How to Strike Energy Efficiency in Food Processing .................. 84
  - 6.6.1 Alternatives in Thermal Food Processing ......................... 84
    - 6.6.1.1 Heat Retrieval .................................................... 84
    - 6.6.1.2 Innovative, Non-Conventional Thermodynamics Cycles .... 85
    - 6.6.1.3 Non-Caloric Processing of Food ....................... 86
    - 6.6.1.4 All-new Heating Techniques ............................. 86
    - 6.6.1.5 Transforming the Wastes into Energy .............. 87
    - 6.6.1.6 Miscellaneous Energy Saving Strategies ......... 87
    - 6.6.1.7 Smart Ways of Food Processing ....................... 87
    - 6.6.1.8 Renewable Energy in Food Processing ............ 88
- 6.7 Roadblocks for Achieving Energy Efficiency in Food Processing Sector ..... 89
  - 6.7.1 Monetary Challenges ......................................................... 89
  - 6.7.2 Technical Challenges ........................................................ 90

DOI: 10.1201/9781003258568-6

6.7.3 Environmental Concerns .................................................................. 90
6.7.4 Legislative and Regulatory Challenges ............................................ 90
6.8 Conclusion................................................................................................. 91
References........................................................................................................ 92

## 6.1 INTRODUCTION

The soaring rates of population growth at a global level have several implications, out of which the rise in food demand is the commonest. For meeting this demand, production of food and food based products needs to be increased to as high as 70% by 2050 in accordance with the estimate of the FAO (The Food and Agriculture Organization), which consequently requires an increase in energy consumption. But the estimated enhancement in the production of energy by the same agency of United Nations is only 30%, thereby clearly indicating an impending imbalance between demand of food and increase in energy generation (Bundschuh et al., 2014; Clairand et al., 2020). This can be rectified by sticking to the objective of being energy efficient in food industries via using less energy for the production of more or at least the same amount of food.

There are multiple divisions in the existing colossal industry of food processing comprising of milk products, oils, confectioneries, fruits, vegetables, liquors, fish, meat and various packed foodstuffs. This industry consumes energy for the purpose of maintaining optimum temperature, better ventilation, lighting, air conditioning, preservation, hygienic plus safe packaging in addition to the conversion of raw materials into processed forms. Out of all of these, regulation of the temperature prevents spoilage of food and beverages besides keeping them fresh for a longer time. Standard ventilation plays an important role in maintaining the hygiene of all the processed food items. Moreover, storage and transportation of processed foods are also counted amongst the energy intensive activities because of their need to consume fossil fuels and electricity.

All these energy guzzling processes and their impending costs stand out somewhere between 20–50% among the total costs of production (Bundschuh et al., 2014). Besides this, out of the total electrical energy consumption in the industrial sector, food industry utilizes a major chunk of about 12% (European Commission, 2015). In contrast to this, energy constraints are known to everyone because of their insufficient production and availability. Also, the multitudes of inefficiencies in food processing methods are responsible for improper consumption of available energy (Degerli et al., 2015; Lin and Xie, 2015). Combustion of fossil fuels during food processing leads to the generation of greenhouse gases such as $CO_2$, $CH_4$, $N_2O$ etc. Such a scenario compels us to look for enhanced energy efficiency in the food processing industry.

The varied components of food processing, if coupled with energy efficient systems, could lead to cost cutting, especially that of the electricity bills. For instance, the processing of food items largely depends on sturdy and efficient techniques adopted for cooling and heating. These two can be integrated to save fuel and reduce greenhouse gas emissions whereby the heat energy generated during the maintenance of the cooling system could be put to use to heat water, simultaneously fixing a gas condenser over the water boiler and transferring this heat to the hydronic heating circuit installed below the floor. All this exercise will result in recycling the same energy for continuous heating and cooling.

On similar lines, lighting systems, if replaced with light emitting diodes (LEDs) in place of fluorescent or incandescent lights, can also contribute to saving energy and the environment by consuming less power in addition to being eco-friendly. Moreover, recycling of waste water and other biodegradable wastes produced as by-products during food processing, also can play a major role in minimizing the energy consumption in this industry. In other words, it can be summed up as, in food processing industry, energy efficiency can be realized at every stage via primarily bringing about a change in the behaviour of parties involved plus subjection to the implementation of internet of things (IoT), which in turn introduces us to the world of novel and efficient technologies. Therefore, the main objective of this chapter is to chalk out the potential of food processing with reference to enhancement of energy efficiency and use.

## 6.2 TRACING THE JOURNEY OF FOOD PROCESSING

Conversion of various food items into processed forms is an age old process comprising of drying in sunlight, adding salts or sugars as preservatives, leavening, adopting cooking methods ranging from steaming, boiling, baking to smouldering. All these not only change the innate natural flavour owing to enzyme catalyzed reactions, but also prevent microbial decay of food items. Preservation of foodstuffs with salt was in vogue primarily in the diets of soldiers and seafarers prior to the advent of knowledge about the technique of canning. Corroborating evidences could be reported from the written records and archaeological remains of the ancient civilizations of Rome, Greece, Egypt, Europe, Asia, North and South America. All the food processing techniques remained more or less the same up to the industrial revolution. Modernization got incorporated into all these methods during the nineteenth and twentieth century in order to meet the needs of soldiers.

It was in the year 1809 when Nicolas Appert invented the technique of hermetic bottling to increase the shelf life of food items for French troops, which with some improvizations became the tinning method and eventually got developed in 1810 into canning with the contributions from Peter Durand (Garcia and Adrian, 2009). Canning techniques got popularized worldwide with the passage of time overcoming the general inhibitions of being expensive and harmful to human health due to the presence of Lead (Pb) in cans. Revolutionary advent of technique of pasteurization in 1864 by Louis Pasteur contributed to the betterment of quality as well as safety of preserved foodstuffs and marked the beginning of preservation of milk along with alcoholic beverages like beer and wine. During the Second World War; the urge to dominate space by overhauling changes in consumption patterns in the twentieth century led to the development of advanced techniques of food processing as the likes of freeze or spray drying, preparation of juice concentrates, evaporation along with additives ranging from artificial sweeteners, colouring agents to preservatives such as sodium benzoate etc. Seeking convenience in every aspect of life resulted in enhancement in the demand of frozen foods, dried instant recipes and reconstituted fruit based drinks in the second half of the twentieth century.

## 6.3 FOOD PROCESSING POTENTIAL

To maintain a healthy economy in most nations, the contribution of the food processing industry is quite significant. According to an estimate, the annual turnover of the processed food and drink industry belonging to the European Union is twice as large compared to that of China and America. The growing demand for food, cost of energy, demographic shifts, changes in regional economies, consumption and urbanization patterns, concerns for food safety and hygiene are some of the major factors affecting the food processing potential worldwide (European Food and Drink Industry, 2013–2014).

The economy of our country is largely dependent on the food processing industry owing to vast range of food products being harvested and processed to fulfil the demands of consumers. Among the different food products, India tops the production of bananas, ginger, guavas, milk, mangoes, okra, papaya, pulses; is second leading producer of wheat, rice, sugarcane, cashew nuts, fish, fruits, poultry, vegetables and tea, while remaining at third place in terms of yield of cereals, cardamom, chicory, coconut, lettuce, mace, nutmeg and pepper on a global level. The increase in incomes coupled with rising demands for good quality packaged food items are the key factors ensuring the sustenance of this industry irrespective of seasonal changes. There is no imminent fear of recession in this industry in addition to having a support system in the form of government initiatives. Market statistics reveal that being one of the largest sectors in the world, this industry will hit an output of US$ 535 billion by 2025–2026 with the generation of 9 million jobs by 2024. Additionally, this sector has so far been able to fetch US$ 4.18 billion in foreign direct investments in the span of six years from 2014–2020. With regard to consumption, India will be in fifth place by 2030 owing to three times the increase in annual utilization in Indian households.

Technological advancements have resulted in making the role of MSMEs quite vital in food processing chains in India. Key growth drivers of this industry can be named as growth of the organized food retail sector, a spurt in online food ordering businesses, a rise in demand for hygienically packed, healthy plus immunity boosting snacks along with government initiatives like the Make in India campaign, Atmanirbhar Bharat etc.; all prioritizing and supporting the sector of food processing to a great extent and transforming our country into a food basket for the entire world.

Besides this, in order to develop the food processing supply chain, the Indian Government has established 18 mega food parks and 134 cold chain projects, which will not only boost the export of food stuffs from this industry but also prevent their wastage. The launch of the GIS One District One Product (ODOP) Digital Map of

India along with Pradhan Mantri Formalization of Micro Food Processing Enterprises Scheme (PM-FME Scheme) on 18 November 2020 by the Indian Ministry of Food Processing have placed the food processing sector on a high growth trajectory (MOFPI, Government of India).

## 6.4 ENERGY CONSUMPTION PATTERNS IN FOOD PROCESSING

A substantial amount of industriousness, machinery and energy goes into how raw food materials are converted into various upscale food products in the food processing sector. Out of these, energy is a key input and varies proportionately in its usage in different nations, though found to be relatively higher in developing countries, as supported by the fact that it stands as high as 55% in certain developing African countries, but in America, it is only around 16% (Sims, 2014). In addition to this, the nature of food items being processed also cast a considerable impact on the energy consumption and energy type. As in the UK fruit and vegetable processing industry, including making French fries, according to an estimate, the fuel and electricity requirements are 13.68MJ/Kg and 1.48MJ/Kg whereas for jam production, these values stand at 1.50MJ/Kg and 0.43MJ/Kg respectively (Ladha-Sabur et al., 2019).

At the consumer end, processed food needs to be refrigerated, heated, ventilated or air conditioned, which again increases the consumption of energy (Wang, 2008). Dependency on different sources of energy further changes according to size or scale of the food industry, which could range from very small sized subsistence level or family unit enterprises, small businesses to large corporate houses exhibiting enhancement of energy consumption with their expansion. Therefore, in order to hit the target of energy efficiency, high energy consuming large food processing industries must be attended to carefully so that the shift in their consumption patterns to become more energy efficient can be actualized.

## 6.5 FOOD PROCESSING: ENERGY CONSUMING TECHNIQUES AT A GLANCE

The major technologies centred around high energy consumption are put to use variously at different steps of food processing which are listed as follows (Clairand et al., 2020):

1. Desiccation or Drying: Harvesting of economic crops such as cereals and pulses is generally followed by this process of drying so that it can be stored and transported safely. The form of energy being utilized for the purpose of attaining correct storage moisture content in these grain crops could vary from LPG (Liquefied petroleum gas), natural gas or simply electricity, estimated to be around 0.5-0.75 MJ/Kg. This step has been found to be amongst the high energy-intensive procedures, particularly in developing countries.

2. Cooking: During this process, there is application of heat to different raw food items and the energy consumption stands at 5–7 MJ/Kg.
3. Parching or Dehydration: To enhance the shelf life of diverse food items, they are subjected to one or the other form of energy so that a considerable reduction in moisture content could be achieved.
4. Evaporation: Some food items are boiled to make them partially or fully free of water, consuming around 2.5–2.7MJ/Kg of energy.
5. Brewing: Processing of a varied range of agricultural produce into beverages and other food items needs energy for either heating or cooling, the amount standing at approx. 50–100 MJ/Kg.
6. Stockpiling or Storage: To maintain the quality and safety of food products, a specific temperature range is required so that these can be stored for longer durations. In general, refrigerators and freezers serve this purpose, but are ranked as high energy consuming processes.

## 6.6 HOW TO STRIKE ENERGY EFFICIENCY IN FOOD PROCESSING

Nowadays, a multitude of options could be cited for achieving the target of energy efficiency in the food processing industry, the only need is their stringent application. A few such energy efficient technologies are summed up here as:

### 6.6.1 Alternatives in Thermal Food Processing

These options are centred around recovery of waste heat, innovative thermodynamics cycles, switching to non-thermal methods and enhancing the reliance upon new heating techniques.

#### *6.6.1.1 Heat Retrieval*

It is well known that in food processing, one of the main energy guzzling processes is heating, however, a part of heat energy is always wasted. This wastage of heat can be rectified by recovering the waste heat by generating power (Hung, 2001). One such technique is the use of heat exchangers as analyzed and validated in the case of milk processing where rescheduling the heat exchangers resulted in a 10 percent increase in energy efficiency (Philipp et al., 2018). Besides this, the waste heat can be recovered by storing it so that it can carry out other processes later on. For example, the refrigeration cycle in a cold chamber was run with the help of photovoltaic energy, which saved the high cost of electricity and a phase change material was made to store in cold temperatures, which can be used in other suitable conditions (Rosiek et al., 2019). However, all these strategies aim at achieving high energy efficiency but their success is largely dependent upon type of fuel used, policy framework and socio-technological structure.

### 6.6.1.2 Innovative, Non-Conventional Thermodynamics Cycles

These include use of all-new energy saving techniques for the purpose of heating and cooling while processing the food items such as

  i. *Heat pipes and heat pumps*: Zero maintenance and cost effective heat pipes making use of latent heat in particular are double sided, one end of them functions as a condenser while the other one as an evaporator. Their heat exchange strategy makes the processing of food, whether in the form of heating or cooling, less time consuming and comparatively easy (34). Heat pumps possess a compressor and valve in addition to the condenser and evaporator of simple heat pipes, making the process of heat exchange being carried out twice by them. The heat pump equipment uses waste heat or any renewable energy form as the heat source which is sent to the evaporator and its heat conditions are made better by the compressor prior to transferring it finally to the condenser (Jouhara, 2018).

  In certain processing methods like pasteurization, direct use of heat pumps can be made, but in others, which have high temperature requirements, they are best utilized to upscale the waste heat of low quality to as high as 150°C, the value asked for by many food industries. Toxicity and flammability of working fluids along with an adequate compressor technology are some of the barriers to the complete success of heat pump operations though they are shown to retrieve waste heat as high as 40% along with cutting down energy costs by 20% (Wang et al., 2018; Frate et al., 2019).

  ii. *Novel cooling cycles*: Unconventional and unique cooling or freezing cycles like adsorption-desorption and ejector refrigeration systems, particularly in small scale milk processing plants analyzed and it was found that the first system uses two separate solutions one for adsorption and the other for refrigeration for transforming low grade heat in heat exchangers. Whereas in the second system, the cycle begins with the vaporization of fluid refrigerant in a boiler, the collection of vapours in an evaporator followed by sufficiently pressurizing it inside the ejector prior to its ultimate transfer into the condenser where it releases heat into the surroundings. This cycle is repeated with the remaining refrigerant fluid, which is first made to enter the boiler and then the valve to lower down the pressure followed by evaporator and subsequently the cooling is made to happen. Ultimately, the vapours of the refrigerant combine with those coming out from boilers and the cycle is re-run (Wang, 2014). Cutting down the consumption of heat by up to 46.1% can be achieved by this method vis à vis traditional methods of freeze drying (Zhang et al., 2018).

  iii. *Combined or hybrid heating techniques*: Diverse sources of energy could be utilized altogether in food processing by implementing a low-temperature based system of hybrid heating, just as was done in the dairy industry where grid electricity, heat pumps in combination with renewable energy, are used for generating heat as an add-on to the other high temperature energy sources (Schumm et al., 2018). However, prior to switching to the implementation of such systems on a large scale, various energy plus innate process constraints are considered thoroughly.

### 6.6.1.3 Non-Caloric Processing of Food

Novel technical strategies yielding the same results but involving no heating at all for the sake of processing food items have now come up, which include: irradiation with high energy electron emitting rays (Beam rays, gamma rays, UV-rays and X-rays) whereby DNA of microorganisms of bacterial and viral nature is degraded, which could otherwise spoil the foodstuffs. This type of sterilization being a cold process not only saves energy costs but also does not alter the organoleptic properties of the food items as observed in pasteurization of milk and disinfection of juice and sausages upon subjecting them to gamma rays, UV rays and beam of electrons respectively (Bhattacharjee et al., 2019; Bouzarjomehri et al., 2020).

Subjection of food items to the oscillations of an electric field for damaging the cell membranes of biological entities (microorganisms in this case) to kill them. This will serve purposes like preserving the physical and nutritional characteristics of food items without increasing the temperature of food or beverages being treated (Gabric et al., 2018), though energy efficiency is not yet completely achieved.

Application of pressure exceeding normal ranges especially to liquid food items, which cause disruption of cell membranes of micro-organismal beings via modifying their lipo-proteinaceous composition. Though this technique is quite advantageous in comparison to conventional methods, particularly with regard to the upkeep of our health, but the increase in temperature of food items is observed and investment cost is also very high (Huang et al., 2020).

Use of membrane assisted filtration in place of conventional separation processes especially in fruit juices and beverages for reaching a specific concentration and clarity. Being performed at low temperatures, this method has a wider applicability as a non-caloric or non-thermal procedure in processing fruit extracts. It saves energy costs, maintains the nutritional composition and quality besides enhancing the production outputs (Bhattacharjee et al., 2019).

### 6.6.1.4 All-new Heating Techniques

These are centred around subjecting the food items to novel and unconventional heating methods for the purpose of sterilizing them and enhancing their shelf life. For example, usage of electromagnetic waves like IR (Infra-red), microwaves and radio frequencies, out of which IR with a high coefficient of heat transfer and low penetration ability has been found most suitable for surface sterilization of food items. Microwaves are less time consuming though cause reduction in moisture content. But, radio frequency heating due to its cost-effectiveness, high penetration ability and more controllable nature is highly preferred over microwave heating (Guo et al., 2019). Further, heating of food items for disinfecting them can also be achieved by the application of direct electric current which is called as Ohmic heating. This method has many merits such as heating of food items in a uniform manner, no thermal disruptions and nutritional alterations in addition to no energy losses because of the conductivity of food material.

### 6.6.1.5 Transforming the Wastes into Energy

Different technical strategies dealing with proper handling of the wastes have been becoming popular of late, out of which their transformation into diverse forms of energy ranging from simple heat to electricity stand out at an important place. With regard to the food industry, this waste to energy transformation approach is of considerable significance because various subsidiary products fertilizers for photosynthesis, SCP (single cell protein) and waste heat are also obtained in addition to generation of electricity (Palanichamy et al., 2002). Food waste can be made to undergo either direct combustion or indirectly subject to thermochemical changes (torrefaction, plasma treatment, pyrolysis or gasification) to form solid fuel like char or liquid fuels like syngas and pyrolysis oil; physicochemical processes (extraction, esterification or re-esterification) to form liquid fuels like vegetable oils or biodiesel and biochemical ones (anaerobic digestion or fermentation) to synthesize liquid fuel like ethanol and gaseous fuel such as biogas, then followed by their combustion to generate thermal energy or electricity.

Out of all of them, production of bio-diesel, bioethanol, biogas or biomass from wastes from agricultural produce such as palm oil, waste cooking oil, fruit lignocellulose, grease trap waste, brewery residues, household organic waste, residue crops, olive mills solid waste, sugarcane bagasse and rice straw etc. have been reported to have similar performance to conventional fuels with fewer emissions.

### 6.6.1.6 Miscellaneous Energy Saving Strategies

Dairy farms can be turned into more energy efficient units by paying attention to areas of concern like air compressing, cooling, pumping and other electric appliances, especially installing renewable sources of energy for water heating which otherwise consumes about 40 percent of the total consumption (Xu and Flapper, 2011; Lima et al., 2018). Further, $CO_2$ emissions could be cut down to 12 percent however, the size of dairy farm has also been to be taken into account (Hosseinzadeh-Bandbafha et al., 2018). Additionally, in the meat industry, about 24 percent of electricity could be saved by installing energy meters for keeping a check on overuse of fuels and other inefficiencies as revealed through energy audits (Ramirez et al., 2006; Nunes et al., 2016; Skunca et al., 2018). Different stages of food processing can also be subjected to various novel approaches for the reduction of usage of electricity, for instance, optimization of food defrosting procedure, sequential ventilation and cold pre-fermentation (especially for wine production), online monitoring mode of energy consumption by different machines along with application of climate control strategies (Damour et al., 2012; Bozchalui et al., 2015; Celorrio et al., 2015; Corrieu et al., 2018).

### 6.6.1.7 Smart Ways of Food Processing

To meet the ever-increasing demands for food (both quantity and quality wise) by the end-users, smart ways of food processing need to be devised. To realize it, the idea of a fourth industrial revolution or Industry 4.0 or Industrial Internet of Things (IIoT) has been rightfully conceived. In it, after being produced in smart factories where skilful manufacturing and distribution occur, the intelligent, sustainable and energy efficient strategies for food processing include computer assisted human interaction, planned handling of logistics, adoption of the state of art technology for the creation of a highly amenable, customized and interconnected production line (Zhou et al., 2015). All the supply chains are expected to be controlled using this smart way of manufacturing with enhanced accuracy by making use of high-tech robotics and knowledge engineering/artificial intelligence (Hasnan & Yusoff, 2018). All the energy wastage could be easily tracked down and dealt with effectively by in time implementation of possible solutions (Kang et al., 2016). Moreover, any impairment in any machinery while processing of food takes place can be easily diagnosed and rectified to prevent the subsequent losses (Mohamed et al., 2019).

### *6.6.1.8 Renewable Energy in Food Processing*

Share of renewable energy is on the rise at the moment on account of the concept of sustainable development where we have to save the natural resources for our future generations as well without compromising our current needs. Application of renewable energy in the food processing industry is one extended aspect of this motive. Photoelectric solar cells/ photovoltaic cells and aero-generators have been proven to be apt substitutes for conventional supply chains of electricity or diesel engine operated generators. Corroborating evidence came from the study conducted in dairy farms in Algeria where a considerable backup of energy as high as 136 GWh/year and assuaging 80 million tons of carbon dioxide via simply introducing the renewable energy resources have been reported, indicating its technical and economic practicality (Nacer et al., 2016; Maammeur et al., 2017).

Energy from solar radiations has been directly used for drying purposes in food processing since ancient times especially in rural areas, however, this process is a bit time consuming, difficult to regulate and keep safe from getting contaminated. Of late, solar food dryers have come up as an amicable solution to all these problems, as substantiated from an analytical research carried out while drying ghost chillies and ginger slices (Rabha et al., 2017). As a major chunk of developing countries are located either near to or within the confines of the equatorial belt, where solar radiations are quite intense, so they can easily switch over to the adoption of solar dryers (Eswara and Ramakrishanarao, 2013).

Further, in Isparta, Turkey's renewable energy system (parabolic to solar collector (PTSC) system) was created to replace the existent typical power system/grid electricity for heating and cooling an ice cream factory. The energy savings from the installation of this PTSC system were up to 98.56% vis-à-vis a conventional system. However, varying ranges of solar radiations, expensive infrastructure and long payback time duration are some of the disadvantages of a PTSC system, which can be eliminated by its unification with other renewable energy alternatives like a trigeneration system. In such systems, photo-electric solar cells/photovoltaic(PV) cells are integrated with combined or hybrid cooling, heating, and power systems and together designated as PV-CCHP or PV-cogeneration/trigeneration (PV-T) systems (Immovilli et al., 2008; Nosrat & Pearce, 2011). For heating and cooling being an integral part of food industries, installation of such systems can revolutionize the energy efficiency measures, but the net profit generation from their incorporation might take longer than usual, as high as 25 years (Basrawi et al., 2014).

## 6.7 ROADBLOCKS FOR ACHIEVING ENERGY EFFICIENCY IN FOOD PROCESSING SECTOR

The different aspects of the food processing industry have their reinforcement in the form of energy consumption, but issues like the cost of production, exhaustion of fossil fuel resources along with adverse impact on environment still remain unaddressed. A whole array of challenges both at technological and non-technological levels poses a roadblock in the path of energy conservation and efficiency in this industry. At the technological level, hurdles in achieving energy efficiency are generally related to intricacies of food processing where the intertwined chains of various procedures occur, casting a long-lasting impact on the quality attributes of end products especially their efficacy and safety. In other words, minimizing the consumption of energy has to be achieved by setting an optimal temperature range and retrieving waste heat along with maintaining the standards of the quality of the final product, though the task is not an easy one. Besides this, the environmental issues arising from using non-renewable energy resources, unsafe disposal of waste products and mismanagement of space or logistics can also not be overlooked. On the other hand, the non-technical blockades comprise of constraints with reference to knowledge and skills dealing with different operators in the food sector essential for saving energy. Here is a brief general account of all these core challenges, which happen upon the path leading to energy efficiency in the food processing industry:

### 6.7.1 Monetary Challenges

Since the attitude of industrialists towards achieving energy efficiency has been found to be quite reluctant owing to the fear of monetary risks involved in terms of implementing possible energy saving solutions. For instance, an additional increase in cost of infrastructure and production amid growing market competition could be very easily accounted for as a hurdle. Moreover, the size of the food industry along with the nature of the products to be processed also need to be taken into consideration while addressing these economic concerns. Though long-term profits are obviously there short term gains are not visible to the entrepreneur. However, this problem could be solved by different economic incentives on the part of Government because otherwise the pertinent market competition among food processing companies would not allow them to stick to energy saving measures (Sims, 2014). A number of existing incentives fall short of their potential for persuading food industrialists to modify their setup for bringing about more energy saving vis-à-vis its consumption because the cost of being energy efficient exceeds.

### 6.7.2 Technical Challenges

Being a technology driven industry, food processing has a huge energy saving potential with regard to the use of heat energy, electrical energy and water. As far as application of thermal energy is concerned, energy can be saved by making use of improvised equipment and processes run inside them. More practical and global solutions must be sought after. For instance, instead of using a gas boiler for generating heat and an air-water chiller to produce frigid temperatures individually, an energy efficient water-water thermal machine capable of generating heat and low temperature simultaneously should be used. In the case of usage of electric power supply, improvisation of equipment must be followed by the maintenance of proper records of energy consumption pattern so that a detailed analysis may suggest other ways to save more energy. In addition to this, reuse and recycling of waste materials plus energy should also be there. However, the complexity of some of the food processing methods does not allow them to be changed to become an all and all energy saviour because this may adversely affect the quality and texture of the final product.

### 6.7.3 Environmental Concerns

The agro-food based processing industry has quite an obvious connection with the environment related issues such as release of greenhouse gases, where it is ranked second in doing so (Qiao et al., 2019). However, energy efficient procedures lower the discharge of greenhouse gases, but the challenge lies ahead in the form of linking renewable energy and other energy saving technologies with this agricultural food processing industry. Slorach et al. in 2020 found in their analytical study focusing on reducing the impact of food waste on the environment in the United Kingdom that the best method was cutting the generation of food waste as compared to the installation of new innovative technologies.

### 6.7.4 Legislative and Regulatory Challenges

Legal issues centred around permissions and regulations are also to be included as possible barriers when it comes to the incorporation of unique, complex technical know-how related to transformation of biomass wastes into varied forms of reusable energies. The developing countries are considered to be the worst sufferers of these barriers in switching over to energy efficient initiatives. Regulations on the level of electric companies of local/regional nature are largely found to be responsible for limiting the installation of industry 4.0 implements, any renewable energy or combined heat and power generation (CHP) plants or commissioning of different demand response (DR) programs. For instance, the implementation of such programmes is quite troublesome in the countries that lag behind in having an organized electricity sector. Additionally, special measures are still to be adopted for bridging the communication gap between government and food industries so that the process of implementing energy efficient strategies could be facilitated. Owing to the fact that, some of the food industries operate only on a small scale, the inclusion of

# POTENTIAL FOR ENERGY EFFICIENCY AND USE

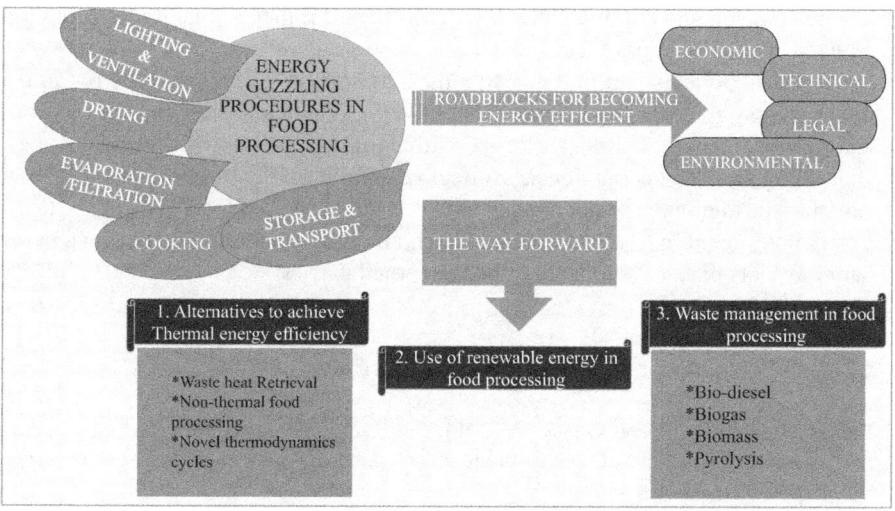

**Figure 6.1** Summing up the problems and solutions with reference to attaining energy efficiency in the food processing industry.

new entities in the energy market like aggregators in the DR sector becomes quite essential (Shoreh et al., 2016).

This entire discussion revolving around food processing potential in terms of energy efficiency and use can be summarized in the Figure 6.1.

## 6.8 CONCLUSION

Various facets related to our lives at the moment like the increase in health problems, the current COVID-19 pandemic, busy lifestyles and the rise in food adulteration, have transformed the buying behaviour of consumers, which revolves around mainly the demand for ready-to-cook/ready-to-eat meals along with healthy, immunity boosting snacks. For instance, we all have observed the rampant growth in the food items belonging to safe and processed food categories such as biscuits and snacks especially amid this COVID-19 crisis.

On the other hand, in terms of consumption of energy and emission of greenhouse gases, the food processing industry has earned an eminent place. Since the proportions of growth of energy competency are not in sync with growth of food demand, future challenges in the field of food products have to be rightfully assessed. That is why, endorsement and adoption of pertinent measures for implementing energy efficient strategies are truly the need of the hour. Different challenges lie ahead for completely transforming the food processing industry for becoming energy efficient, however, novel and innovative measures like waste heat retrieval, energy storage options, enhanced reliance upon renewable energy resources, application of non-thermal food processing techniques, biomass waste management can be counted among some up and coming pathways in this regard if their actual implementation is realized at the grass root level.

Befitting policies and initiatives on the administrative level must be put in place to inspire entrepreneurs to resort to these energy efficiency strategies. Their worries and discomforts for building a proper infrastructure along with adoption of relevant procedures to become energy effective should be dealt with using the utmost care. For abridging the widening gap between production of food and energy market, timely investment in terms of financial, technological, environmental, legal and regulatory aspects have to be made by the concerned parties.

## REFERENCES

Basrawi F, Yamada T, and Obara SY. 2014. Economic and environmental based operation strategies of a hybrid photovoltaic–microgas turbine trigeneration system. *Applied Energy* 121: 174–183.

Bhattacharjee C, Saxena VK, and Dutta S. 2019. Novel thermal and non-thermal processing of watermelon juice. *Trends in Food Science & Technology* 93: 234–243.

Bouzarjomehri F, Dad V, Hajimohammadi B, Shirmardi SP, and Salimi AY. 2020. The effect of electron-beam irradiation on microbiological properties and sensory characteristics of sausages. *Radiation Physics and Chemistry* 168: 108524.

Bozchalui MC, Cañizares CA, and Bhattacharya K. 2014. Optimal operation of climate control systems of produce storage facilities in smart grids. *IEEE Transactions on Smart Grid* 6(1): 351–359.

Bundschuh J, Chen G, and Mushtaq S. 2014. Towards a sustainable energy technologies based agriculture. *Sustain Energy Solut Agric* 3(15): 3–15.

Celorrio R, Martínez E, Saenz-Díez JC, Jiménez E, and Blanco, J. 2015. Methodology to decrease the energy demands in wine production using cold pre-fermentation. *Computers and Electronics in Agriculture* 117: 177–185.

Clairand JM, Briceño-León M, Escrivá-Escrivá G, and Pantaleo AM. 2020. Review of energy efficiency technologies in the food industry: trends, barriers, and opportunities. *IEEE Access* 8: 48015–48029.

Corrieu G, Perret B, Kakouri A, Pappas D, and Samelis J. 2018. Positive effects of sequential air ventilation on cooked hard graviera cheese ripening in an industrial ripening room. *J Food Eng* 222: 162–168.

Damour C, Hamdi M, Josset C, Auvity B, and Boillereaux L. 2012. Energy analysis and optimization of a food defrosting system. *Energy* 37(1): 562–570.

de Lima LP, de Deus Ribeiro GB, and Perez R. 2018. The energy mix and energy efficiency analysis for Brazilian dairy industry. *Journal of Cleaner Production* 181: 209–216.

Degerli B, Nazir S, Sorgüven E, Hitzmann B, and Özilgen M. 2015. Assessment of the energy and exergy efficiencies of farm to fork grain cultivation and bread making processes in Turkey and Germany. *Energy* 93: 421–434.

Eswara AR, and Ramakrishnarao M. 2013. Solar energy in food processing—a critical appraisal. *Journal of food science and technology* 50(2): 209–227.

European Commission. Energy Balancesheets 2015 Data Eurostat. Accessed: 2017. [Online]. Available: http://ec.europa.eu/eurostat/ documents/3217494/8113778/KS-EN-17-001-EN-N.pdf/99cc20f1- cb11-4886-80f9-43ce0ab7823c.

Frate GF, Ferrari L, and Desideri U. 2019. Analysis of suitability ranges of high temperature heat pump working fluids. *Applied Thermal Engineering* 150: 628–640.

Gabrić D, Barba F, Roohinejad S, Gharibzahedi SM, Radojčin M, Putnik P, and Bursać Kovačević D. 2018. Pulsed electric fields as an alternative to thermal processing for preservation of nutritive and physicochemical properties of beverages: A review. *Journal of Food Process Engineering* 41(1): e12638.

Garcia R, and Adrian J. 2009. Nicolas Appert: Inventor and manufacturer. *Food Reviews International* 25(2): 115–125.

Guo C, Mujumdar AS, and Zhang M. 2019. New development in radio frequency heating for fresh food processing: A review. *Food Engineering Reviews* 11(1): 29–43.

Hasnan NZN, and Yusoff YM. 2018. Short review: Application areas of industry 4.0 technologies in food processing sector. In 2018 *IEEE Student Conference on Research and Development (SCOReD)* 1–6.

Hosseinzadeh-Bandbafha H, Safarzadeh D, Ahmadi E, and Nabavi-Pelesaraei A. 2018. Optimization of energy consumption of dairy farms using data envelopment analysis–A case study: Qazvin city of Iran. *Journal of the Saudi Society of Agricultural Sciences* 17(3): 217–228.

Huang HW, Hsu CP, and Wang CY. 2020. Healthy expectations of high hydrostatic pressure treatment in food processing industry. *Journal of Food and Drug Analysis* 28(1): 1–13.

Hung TC. 2001. Waste heat recovery of organic Rankine cycle using dry fluids. *Energy Conversion and Management* 42(5): 539–553.

Immovilli F, Bellini A, Bianchini C, and Franceschini G. 2008. Solar trigeneration for residential applications, a feasible alternative to traditional micro-cogeneration and trigeneration plants. *In Proc. IEEE Ind Appl Soc Annu Meeting* 1–8.

Jouhara H, Khordehgah N, Almahmoud S, Delpech B, Chauhan A, and Tassou SA. 2018. Waste heat recovery technologies and applications. *Thermal Science and Engineering Progress* 6: 268–289.

Kang HS, Lee JY, Choi S, Kim H, Park JH, Son JY, Kim BH, and Noh SD. 2016. Smart manufacturing: Past research, present findings, and future directions. *International Journal of Precision Engineering and Manufacturing-Green Technology* 3(1): 111–128.

Ladha-Sabur A, Bakalis S, Fryer PJ, and Lopez-Quiroga E. 2019. Mapping energy consumption in food manufacturing. *Trends in Food Science & Technology* 86: 270–280.

Lin B, and Xie X. 2015. Factor substitution and rebound effect in China's food industry. *Energy Conversion and Management* 105: 20–29.

Maammeur H, Hamidat A, Loukarfi L, Missoum M, Abdeladim K, and Nacer T. 2017. Performance investigation of grid-connected PV systems for family farms: case study of North-West of Algeria. *Renewable and Sustainable Energy Reviews* 78: 1208–1220.

Mohamed N, Al-Jaroodi J, and Lazarova-Molnar S. 2019. Leveraging the capabilities of industry 4.0 for improving energy efficiency in smart factories. *Ieee Access* 7: 18008–18020.

Nacer T, Hamidat A, and Nadjemi O. 2016. A comprehensive method to assess the feasibility of renewable energy on Algerian dairy farms. *J Cleaner Prod* 112: 3631–3642.

Nosrat A, and Pearce JM. 2011. Dispatch strategy and model for hybrid photovoltaic and trigeneration power systems. *Applied Energy* 88(9): 3270–3276.

Nunes J, Silva PD, Andrade LP, and Gaspar PD. 2016. Key points on the energy sustainable development of the food industry–Case study of the Portuguese sausages industry. *Renewable and Sustainable Energy Reviews* 57: 393–411.

Palanichamy C, Babu NS, and Nadarajan C. 2002. Municipal solid waste fueled power generation for India. *IEEE Transactions on Energy Conversion* 17(4): 556–563.

Philipp M, Schumm G, Heck P, Schlosser F, Peesel RH, Walmsley TG, and Atkins MJ. 2018. Increasing energy efficiency of milk product batch sterilisation. *Energy* 164: 995–1010.

Qiao H, Zheng F, Jiang H, and Dong K. 2019. The greenhouse effect of the agriculture-economic growth-renewable energy nexus: evidence from G20 countries. *Science of the Total Environment* 671: 722–731.

Rabha DK, Muthukumar P, and Somayaji C. 2017. Energy and exergy analyses of the solar drying processes of ghost chilli pepper and ginger. *Renewable Energy* 105: 764–773.

Ramirez CA, Patel M, and Blok K. 2006. How much energy to process one pound of meat? A comparison of energy use and specific energy consumption in the meat industry of four European countries. *Energy* 31(12): 2047–2063.

Rosiek S, Romero-Cano MS, Puertas AM, and Batlles FJ. 2019. Industrial food chamber cooling and power system integrated with renewable energy as an example of power grid sustainability improvement. *Renewable Energy* 138: 697–708.

Schumm G, Philipp M, Schlosser F, Hesselbach J, Walmsley TG, and Atkins MJ. 2018. Hybrid heating system for increased energy efficiency and flexible control of low temperature heat. *Energy Efficiency* 11(5): 1117–1133.

Shoreh MH, Siano P, Shafie-khah M, Loia V, and Catalão JP. 2016. A survey of industrial applications of Demand Response. *Electric Power Systems Research* 141: 31–49.

Sims RE, Bundschuh J, and Chen G. 2014. Global energy resources, supply and demand, energy security and on-farm energy efficiency. *Sustainable Energy Solutions in Agriculture* 1: 19–52.

Skunca D, Tomasevic I, Nastasijevic I, Tomovic V, and Djekic I. 2018. Life cycle assessment of the chicken meat chain. *Journal of Cleaner Production* 184: 440–450.

Slorach PC, Jeswani HK, Cuéllar-Franca R, and Azapagic A. 2020. Assessing the economic and environmental sustainability of household food waste management in the UK: Current situation and future scenarios. *Science of The Total Environment* 710: 135580.

Wang JF, Brown C, and Cleland DJ. 2018. Heat pump heat recovery options for food industry dryers. *International Journal of Refrigeration* 86: 48–55.

Wang L. 2008. *Energy efficiency and management in food processing facilities*. CRC press, 117–128.

Wang L. 2014. Energy efficiency technologies for sustainable food processing. *Energy efficiency* 7(5): 791–810.

Xu T, and Flapper J. 2011. Reduce energy use and greenhouse gas emissions from global dairy processing facilities. *Energy policy* 39(1): 234–247.

Zhang S, Luo J, Wang Q, and Chen G. 2018. Step utilization of energy with ejector in a heat driven freeze drying system. *Energy* 164: 734–744.

Zhou K, Liu T, and Zhou L. 2015. Industry 4.0: Towards future industrial opportunities and challenges. In *2015 12th International conference on fuzzy systems and knowledge discovery (FSKD)* 2147–2152.

CHAPTER 7

# Global Food Security and Effects of Various Environmental Constraints on Food Crops

Tunisha Verma,[1#] Sahima Tabasum,[2#] Savita Bhardwaj,[1#] Vandana Gautam,[3] Bharat Kapoor,[4] and Dhriti Kapoor[1*]

[1]Department of Botany, School of Bioengineering and Biosciences, Lovely Professional University, Phagwara (Punjab), India
[2]Department of Chemistry, School of Chemical Engineering and Physical Sciences, Lovely Professional University, Phagwara (Punjab), India
[3]College of Horticulture and Forestry, Dr. Y. S. Parmar University of Horticulture and Forestry, Nauni, Solan (HP), Neri Campus (Himachal Pradesh), India
[4]Department of Hotel Management and Tourism, Guru Nanak Dev University, Amritsar (Punjab), India
*Corresponding author email: dhriti405@gmail.com
#Equal Contribution

## CONTENTS

| | | |
|---|---|---|
| 7.1 | Introduction | 96 |
| 7.2 | Current Scenario of Global Food Industry | 96 |
| 7.3 | Causes of Global Food Waste | 99 |
| 7.4 | Effects of Various Abiotic Stresses on Food Crops | 101 |
| | 7.4.1 Salinity Stress | 102 |
| | 7.4.2 Drought Stress | 102 |
| | 7.4.3 Temperature Stress | 103 |
| | 7.4.4 Metal Stress | 104 |
| 7.5 | Methods to Improve Food Security | 104 |
| 7.6 | Conclusion | 105 |
| References | | 105 |

DOI: 10.1201/9781003258568-7

## 7.1 INTRODUCTION

Around 800 million and 1.2 billion people are suffering from hunger and malnutrition worldwide over the last 40 years (Gibson, 2012). Currently, the world population is increasing at an alarming rate and it is difficult for the current system of food production to provide food for over 9 billion people. Around 70% increase in food demand is estimated for the present food manufacture. Over exploitation of agricultural land has occurred due to intensive agriculture practices, which cause alterations in the climatic change. Change in the ecological conditions resulted in the effect on the environment, hence immense research is needed to develop novel and sustainable techniques to increase food production (Leandro et al., 2020; Abiad and Meho, 2018; Santeramo and Lamonaca, 2021). Deficiency of both macro and micro nutrients and unbalanced diet are also responsible for food insecurity (Santeramo and Shabnam, 2015; Shabnam et al., 2016).

Food nutrition and security is also attained through the food science and technology via using various technologies such as food processing, preservation, management etc. to extend the shelf life of products (Martindale, 2017; Augustin et al., 2016). Loss of fresh produce can be ignored by using good post-harvest handling practices from farm to retail. This aspect is fetching high attention as the food has to be provided to the increasing population in the urban areas and towns that is produced in rural parts. To ensure appropriate food supply, proper transportation, organizational facilities, better preservation and management practices are required (Cole et al., 2018). Food waste byproducts that are released from food processing can be alternatively used for non food purposes such as composting, animal feed and for the production of chemicals (Cole et al., 2018). There is complexity in the nutritional food security because on one side we are supposed to improve the quantity of accessible food while on the other side more than 2 billion people are overweighted.

Food security can be significantly improved without causing any adverse effect on the environment by decreasing the over consumption in this population. Proper dietary guidelines and change in consumer behaviour through education is required so that everyone can get healthier processed food (Lewis and Burton-Freeman, 2010). Innovative technologies should be used to build future developments around sustainable food supply networks, which can only be accomplished through innovative value adding techniques in an organizational learning environment. This technique will be helpful to improve food security and to provide healthier food to improve the quality of our lives (McCarthy et al., 2013). Adoption of resource efficiency is one such significant strategy to achieve global food security which aims to provide sustainable crop production across the worldwide. Hunger and malnutrition can be effectively reduced through agricultural growth (Devaux et al., 2014).

## 7.2 CURRENT SCENARIO OF GLOBAL FOOD INDUSTRY

The word "food industries" refers to a set of industrial operations involving the production, distribution, processing, conversion, preparation, preservation, transportation, certification, and packing of food. The food business has evolved into a highly diverse industry, with manufacturing ranging from small, labor-intensive, family-run operations to huge, capital-intensive, highly automated industrial processes. Many food-related enterprises are nearly wholly reliant on local agriculture, vegetables and fisheries. The food industry includes: agriculture raising crops, livestock and seafood and agrichemicals, agricultural construction, firm machinery and supplies etc. Food processing includes both the preparation of fresh items for market and the production of ready-to-eat foods. In marketing general product promotion includes milk board, new product development, advertising, marketing campaigns, packaging, public relations and in logistics transportation and warehousing in wholesale and food distribution (Mc Carthy et al., 2018).

Grocery includes farmers' markets, public markets and other retailers' food services (which include catering). Regulation includes laws and regulations for food production and sale at the local, regional, national and international levels including food quality, food security, food safety, marketing/advertising and industry lobbying (Stephens et al., 2017). The majority of food produced for the food business is derived from commodity crops grown using traditional agricultural methods. Agricultural is the practice of cultivating particular plant and rearing domesticated animals to produce food, feed goods, fiber, and other desirable items (livestock). On average, terrestrial agriculture produces 83% of food consumed by people. Aquaculture and fishing are two other food sources. Agriculture also includes scientists, inventors and others who work to improve agricultural practices and tools. Agriculture employs one out of 10, which is about 4% of national GDPs on average in 2017 (Stephens et al., 2017).

Agriculture accounts between 14 and 28% of global greenhouse gas emissions, making it one of the most significant contributors yet it only accounts for 3% of global GDP (Lobell and Gourd, 2012). A sustainable food system provides food security and nutrition for everyone while preserving the economic, social and environmental foundations necessary to provide food security and nutrition for future generations. This implies that it is lucrative throughout, guaranteeing economic sustainability; as it has widespread societal benefits, ensuring social sustainability; and has a positive or neutral influence on natural resources, ensuring environmental sustainability (FAO, 2015). Climate change is widely considered to be a major threat to future food security. Although the exact consequences of climate change are impossible to predict, the general view is that global crop production will be negatively affected (ERS, 2017; World Economic Forum, 2017).

The worsening effects of economic shocks, especially those originating from the COVID-19 epidemic, created the biggest global economic crises since World War I, hitting impoverished nations disproportionately and increasing already unstable situations, notably in countries where wars are occurring. As a result of significant job and income losses, tens of millions of vulnerable individuals were unable to buy enough food, which was often accompanied by rapidly growing and persistent high food costs. Economic shocks (including those caused by COVID-19) were considered the primary driver of acute food insecurity in 17 countries in 2020, accounting for over 40 million people in Crises or worse (IPC/CH Phase 3 higher) or equivalent, compared to eight countries in 2019 with around 24 million people (World Food Program, 2021).

While the forecast for global food supply remains positive, food prices be at an all-time high as a result of rising input costs, which when combined with high transportation costs, are driving up import expenses. This has the greatest impact on impoverished and developing countries, as they rely on food imports the most. The Agricultural Commodity Price Index is 35% higher in February 2022 than it was in January 2021. Maize and wheat prices are 26% and 23% higher, respectively, than they were in January 2021, while rice prices are around 17% lower.

The picture for food and nutrition security in many low- and middle-income nations is bleak, with economic recovery faltering, COVID-19 uncertainties and disruptions persisting, and fiscal capacity deteriorating. According to a UN report on the state of Food Security and Nutrition in the World, between 720 and 811 million people would go hungry in 2020. Around 118 million more individuals faced chronic hunger in 2020 than in 2019, based on the center of the anticipated range (768). According to a separate statistic that analyses year-round access to enough food, approximately 2.37 billion people (or 30% of world population) lacked appropriate food in 2020, up 320 million from the previous year (World Food Program, 2021).

There is an increase in the number of people facing acute food insecurity in 2020–2021 according to recent studies. Acute food insecurities are described as a situation in which a person's life or livelihood is jeopardized due to a shortage of food. According to the Global Network Against Food Crises, 161 million people in 2021 suffered "crisis" acute food insecurity in 2021, over 7% higher than the previous year (Food Security and the World Bank).

The global agri-food business is in constant flux and restructuring in order to fulfil the ever-changing nutritional requirements and preferences of customers throughout the world. The endeavor to strike a trade-off between product price vs safety, quality, diversity, and demand is one of the industry's key difficulties. Industrial practitioners must handle each of the primary important areas in which a product is created, processed, stored, distributed, and accessible throughout the world to successfully meet these problems and achieve cost savings. Similarly, recent studies have reported that, in order to achieve market success, global food producing industries are now adopting strategies aimed at gaining a competitive advantage through category focus as opposed to other industries where the key players focus on portfolio management (in this $4 trillion per year industry) (United States Department of Agriculture, 2016).

Adopting a product-focused approach allows businesses to become actual leaders in their fields for specific goods while also reaching global economies of scale. Agriculture and agriculture-related businesses in the United States contributed $835 billion to the country's gross domestic product (GDP) in 2014.

The agricultural production of the United States contributed $177.2 billion to this total, or nearly 1% of GDP. Agriculture employed 17.3 million full- and part-time workers in 2014, accounting for 9.3% of total employment in the United States. Over 2.6 million of these occupations were supported by direct on-farm employment, with an additional 14.7 million supported by inter-related industries. Food services and food/beverage manufacturing supported 1.8 million jobs, whereas food/

beverage manufacturing supported 11.4 million jobs. In 2013, the food and beverage manufacturing industry in the United States employed roughly 1.5 million people, accounting for all nonfarm employment in country (USDA, 2016). Other countries are seeing similar patterns.

For example, the agricultural and agri-food sector generated $106.9 billion in 2013, accounting for 6.7% of Canada's overall GDP – a trend that has continued every year since 2007, with the exception of the 2009 economic slump. The majority of industries in this sector continued to grow, and employed over 2.2 million people, accounting for one out of every eight jobs in Canada. In the AAFS, the food service business was the greatest employer, accounting for 5.3% of all occupations in Canada. On the opposite side of the Atlantic, the European Union bought close to €60 billion in agricultural goods from developing nations between 2008 and 2010.

In Europe, the food value chain generates €800 billion in added value and €4 trillion in revenue. In agriculture, the food industry, food trade, and services, 46 million people are employed in almost 15 million holdings or companies. In terms of turnover (€1.22, or 1.8% of EU Gross value Added-GVA). Employment (4.22 million jobs), value added (€206 billion, or 12.8% of EU manufacturing), and exports (€ 92 billion, or 18% of EU exports), the food and beverages industry is the EU's largest manufacturing sector. Small and Medium Enterprise (SME) businesses make up 99.1% of the industry. Regardless of geographic location, the contribution of the family farmer in alleviating world hunger has received a lot of attention (World Bank, 2010; Asian Development Bank, 2013).

There are up to 500 million family farms in the globe, accounting for 98% of all agricultural holdings. Fruit and vegetable farms, grain farms, orchards, cattle ranches, and even fisheries and those that collect non-wood forest products are examples of typical family farms. Family farms contribute up to 40% of main crops in Brazil and 84% of all produce in the United States. Many people see such farmers as caretakers of the land rather than exploiters since they have extensive awareness of their land's history, requirements, and agricultural capabilities. All this data reminds us of the importance of global food industry and security, as well as the influence it has on key global economies (FAO, WFP, and IFAD, 2012).

## 7.3 CAUSES OF GLOBAL FOOD WASTE

Food wastage and loss are a major concern in efforts to alleviate hunger, increase income, and enhance food security in the world's poorest countries. Food losses have an impact on impoverished people's food security, food quality and safety, economic development, and environmental protection. The exact reasons why food losses differ around the world and are highly reliant on the individual circumstances and local scenario in each country. Crop production decisions and patterns, internal infrastructure and capacity, marketing chains and distribution networks, and consumer purchasing and food usage practices will all influence food losses in broad terms. Food losses should be kept to a minimum regardless of a country's level of economic growth or system maturity. Food losses are a waste of resources such as land, water, energy and inputs utilized in production. Producing food that will not be consumed results in wasteful $CO_2$ emissions as well as a loss of the food's economic worth.

From agricultural production to final consumption purposes, food is lost or wasted along the supply chain. Food is wasted to a large extent during the consumption stage in middle- and high-income nations, meaning it is rejected even through it is still safe for human consumption. In the industrialized world, significant losses occur early in the food supply chain. Food is lost largely in the early and intermediate stages of the food supply chain in low-income nations; significantly less food is wasted at the consumer level. On a per-capita basis, the industrialized world wastes significantly more food than developing countries. We estimate that per capita food waste in Europe and North America is 95–115 kg per year, while it is just 6–11 kg per year in sub-Saharan Africa and South/Southeast Asia (ERS, 2017; World Economic Forum, 2017).

Food losses and waste are mostly caused by financial, managerial, and technical limits in harvesting procedures, storage and cooling facilities in severe climatic conditions, infrastructure, packing and marketing systems in low-income nations. Given that many smallholder farmers in developing nations are living on the edge of food poverty, a reduction in food losses could have a large and immediate impact on their livelihoods. Food supply chains in developing nations must be enhanced by enabling small farmers to organize, diversify, and scale up their production and marketing, among other things. Infrastructure, transportation, the food industry, and the packing industry all demand investment (World Food Program, 2021). In order to achieve this, both the public and private sectors must work together. Food losses and waste in middle- and high-income nations are mostly caused by consumer behavior as well as a lack of cooperation among supply chain partners. Sales agreements between farmers and buyers may lead to farm crops being wasted. Food can be thrown out because of quality standards that reject food that isn't in perfect condition, aesthetically or otherwise. Inadequate purchase planning and expiring "best-before" dates, along with the reckless attitude of those consumers who can afford to waste food, result in enormous volumes of waste at the consumer level. Food waste can be minimized in developed countries by improving awareness among the food industry, merchants, and consumers. There is a need to find a good and beneficial use for safe food that is presently thrown away (Global Network Against Food Crises, 2021).

In US every year, 30% of all food is thrown away, costing US$48.3 billion (€32.5 billion). Since agriculture is the greatest human use of water, it is estimated that nearly half of the water required to create this food is wasted. Farm losses are likely to range between 15 and 35%, depending on the industry. The retail industry has a rather high loss rate of over 26%, although supermarkets, unexpectedly, only lose approximately 1%. Annual losses are estimated to be in the range of US$90 billion to US $100 billion (Lundqvist et al., 2008). In Latin America, the food that is presently lost or squandered could feed 300 million people (FAO, 2011).

In Europe every year, families in the United Kingdom waste an estimated 6.7 million tons of food, accounting for around one-third of the 21.7 million tons purchased. This means that over 32% of the food purchased each year is not consumed. Local governments presently collect the majority of this (5.9 million tons, or 88%). The majority of food waste (4.1 million tons, or 61%) could have been avoided and eaten if it had been better handled (Knight and Davis, 2007). Food that is now being thrown away in Europe could feed millions of people (FAO, 2012).

In a study of more than 1600 Australian families conducted for the Australia Institute in 2004, it was discovered that $10.5 billion was spent on products that were never used or thrown away. This equates to more than $5,000 per person each year. In Africa food waste is caused by inefficient processing and drying, inadequate storage, and a lack of infrastructure. Post-harvest food losses in Sub-Saharan Africa are estimated to be worth $4 billion each year, enough to feed at least 48 million people. Post-harvest losses of food cereals are reported to be 25% of the entire crop produced in several African nations. Post-harvest losses can approach 50% for some crops, such as fruits, vegetables, and root crops, which are less resilient than cereals. Economic losses in the dairy business in East Africa might be as high as US$90 million per year owing to spoiling and wastage (FAO). Every year, almost 95 million liters of milk worth $22.4 million are wasted in Kenya.

Each year, Tanzania loses roughly 59.5 million liters milk, accounting for about 16% of total dairy output during the dry season and 25% during the rainy season. Approximately 27% of all milk produced in Uganda is lost, equating to US$23 million each year. Africa's current food loss could feed 300 million people (Food and Agriculture Organization, 2011). In Asian countries like China wastes 50 million tons of grain per year, accounting for one-tenth of the country's total grain output, according to statistics. It is also estimated that enough food is wasted each year to feed 200 million people, or nearly one-sixth of the country's population. According to the Food Corporation of India, losses for grains and oilseeds are smaller, ranging between 10 and 12%. Each year, 23 million tons of food cereals, 12 million tons of fruits, and 21 million tons of vegetables are lost, totaling 240 billion rupees in value both in harvest and post-harvest time. According to the Ministry of Food Processing, India wastes agricultural output worth 580 billion rupees each year (Knight and Davis, 2007).

## 7.4 EFFECTS OF VARIOUS ABIOTIC STRESSES ON FOOD CROPS

Global food security warrants the need of food to keep pace with the growing demand of the population. To address food supply deficiency, serious measures are required to reach predicted growth rates in order to assure consistent supply of food, particularly when agriculture land loses fertile lands due to urbanization and industrialization. Furthermore, previous efforts to improve agricultural production in order to satisfy the rising need of food were influenced by degradation of land and the impact of climate variability, increased the components, frequency, and occurrence of abiotic stresses (Qadir et al., 2014). The principal element limiting agricultural production is environmental stress. Plants subjected to a variety of abiotic stresses such as drought, frost, freezing, salt, metal, mineral toxicity, and nutrient deficiencies which play an important role in determining yield, biomass, and dispersion of different plant species in diverse environments. These abiotic stresses limit plant growth and development, resulting in crop output ranging from 50% to 70% (Thakur et al., 2010).

### 7.4.1 Salinity Stress

Soil salinity is a serious issue in arid and semiarid climate, where plants evaporate and transpire at a faster rate. It arises due to poor management of agricultural practices, lack of proper drainage system, drought, excessive evaporation and transpiration (Flowers, 2004). Soil with a higher concentration of salts in it is called saline soil. Thus, an increase in salt content is known as soil salinity or salinization (Bockheim, 2000). Salinity can be induced naturally due to the presence of salts in the subsoil (primary salinity), or it can be introduced into the soil as a result of increased anthropogenic activity like excessive use of fertilizer or saline water for irrigation purposes resulting in environmental degradation (secondary salinity) (Carillo, 2011). Higher salt accumulation poses a serious threat to the agriculture industry as it severely affects growth, yield, quality, and development of seed by reducing crop output. Salinity causes stress in plants by disrupting ionic and osmotic equilibrium, leading to osmotic stress. Osmotic stress is caused by higher accumulation of salt, which reduces the quantity of water that plants utilize, culminating physiological drought. As a result of the combined effects of greater osmotic potential and specific ion toxicity, salt stress affects plant quality and quantity by slowing seed germination, impairing growth and development phases.

Salinity-induced stress has a negative impact on plant development, membrane function, cytosolic metabolism, ROS production and increases cell senescence during extended exposure. Growth inhibition is the major injury resulting in various symptoms, which ultimately leads to cell death under serious salinity shock. Excessive accumulation of salts result in nutrient disturbances by altering accessibility, transport and partitioning of nutrients or imbalances due to comparison of $Na^+$ and $Cl^-$ with other nutrients. Specific ion toxicity of $Na^+$ and $Cl^-$, as well as ionic imbalances, influence biophysical components and/or metabolism of plant growth in saline conditions (Akbari et al., 2007).

### 7.4.2 Drought Stress

Plants are exposed to different abiotic stresses during their growth and development under natural conditions. Drought is the most severe abiotic stress that affects plant development and output significantly. It is a climate term that describes a period of time with less rainfall and is defined as stress connected to a water deficit. Water makes up approximately 80% of the biomass. It is vital component of plant survival and is required for nutrient transfer. As a result, lack of water causes drought stress and results in plants losing their vigour (Ashkavand, 2018). Plants experience osmotic stress as a result of higher level of salts. Elevated levels of salt conditions reduces soil water potential as osmotic potential of salt is less than water, making it hard for the roots to absorb water in the soil. Owing to enhanced loss of water via transpiration or evaporation and temperature variation trigger drought stress (Boyer, 1982). Rise in the level of toxic ions in the cells causes potential damage in the plant due to drought stress. Shortage of water affects plant growth and development that leads to decrease in plant yield. Drought is harmful because it causes cellular water outflow, which leads to plasmolysis and cell death (Harris et al., 2002). Scarcity of water disrupts plant growth and development reducing plant yield and significantly affects germination of seed (Kaya et al., 2006). Research done on pea varieties exhibited that drought stress significantly hampered the germination and early vegetative growth (Okcu et al., 2005). Water deficit conditions generated by polyethylene glycol reduced germination potential, fresh and dry weight of shoot and root and hypocotyl length in *Medicago sativa* (alfalfa) (Zeid and Shedeed, 2006). Likewise in rice, drought stress during the vegetative stage greatly reduced the plant growth and development (Tripathy et al., 2000; Manikavelu et al., 2006).

## 7.4.3 Temperature Stress

Plant scientists all across the world are concerned about changing climatic conditions. Extreme temperature changes are the leading cause of several abiotic stresses, which have a significant impact on the agriculture industry. Variation in temperature, whether positive or negative, has an unfavorable impact on the plant's growth, development, and production. High temperature stress refers to rise in temperature beyond the threshold value for longer periods and has a devastating impact on plant growth and development (Watanabe et al., 2009, Shah et al. 2011). Heat stress results in numerous physiological injuries such as scorching of leaves, inhibitory growth of stems and roots, leaf defoliation, cell death and fruit damage which ultimately lead to loss of a plant's productivity (Vollenweider and Günthardt-Goerg, 2005). High temperatures result in drought stress due to increased water loss by transpiration and evaporation. In many cases, heat stress alters crop morphology, lengthens hypocotyls and petioles stimulating morphological responses to escape shade. Heat stress, on the other hand, has an impact on net shoot assimilation rates and plant total dry weight. It also speeds up sexual development, which cuts down on the time it takes to add photosynthesis to fruit or seed production. Aside from high temperatures, cold temperatures also diminish agricultural output by decreasing quality and shelf life after harvest. Plants suffer from cold stress when they are exposed to extremely low temperatures. It is caused by temperatures that are chilly enough to cause harm without causing ice crystals to form in plant tissues, whereas freezing stress causes ice to form in plant tissues (Xin et al., 2000). Chilling stress has a direct effect on the plant because it alters membrane shape and reduces protoplasmic streaming, causing cellular damage in it (Sanghera et al., 2011).

Cool temperature is enough to produce injury without forming ice crystals in plant tissues, whereas freezing stress results in ice formation within plant tissues (Aroca et al., 2003). Chilling stress has a direct impact on plants as it changes membrane structure and decreases protoplasmic streaming leading to cellular damage. Depending on the severity of the stress, low temperatures have a significant impact on numerous components of crop growth, including cell division, cell metabolism, photosynthesis, respiration rate, and water transport (Farooq et al., 2009). Plant development is harmed as a result of cellular damage, altered metabolism, resulting in improper fruit ripening, internal discoloration (vascular browning), and increased susceptibility to decay, finally leading to the senescence in plant (Nahar et al., 2009).

### 7.4.4 Metal Stress

Metal stress is caused when the level of non-biodegradables, determined inorganic chemicals exceeds the toxic level on cells and genes resulting in mutagenic impacts on crops by contaminating irrigation, soil, drinkable water, food chains and the surrounding environment. Inadequate crop nutrition induces lack of some essential elements, with effects on growth, yield, or quality of the produce depending on the species (Flora et al., 2008). Based on the availability of nutrients metals discovered in soils are considered to be vital micronutrients requiring plant growth (Mg, Zn, Cu, and Ni) and other as non-essential elements with little physiological function (Cr, Co, Pb, and Hg) (Wuana et al., 2011). Excessive use of heavy metal acts as a limiting factor in different urban and agricultural areas. Heavy metal at a toxic level hampers physiological and metabolic functioning in plants by formation of highly reactive bonds between heavy metals and sulfhydryl groups which interfere with the normal functioning of the cellular molecules (Schutzendubel et al., 2002).

## 7.5 METHODS TO IMPROVE FOOD SECURITY

Government policymakers have long placed a high priority on food. However, in the past few decades, agricultural productivity has suffered a major setback as food security has received far less attention than policy goals centered around economic and social development (Beddington et al., 2010) It is as if the Green Revolution's seeming triumphs have persuaded decision-makers that food security will take care of itself (Foresight, 2011).

With an increase in population there has been a rise in demand outweighing supply, reversing the progress made in alleviating hunger globally (Wheeler et al., 2013). Challenges in the food system like global climatic change, average increase in purchasing power, resource scarcity and consequent diet change make it difficult to maintain the same rate of productivity in the future.

A previous framing of the food security solution suggested that taking advantage of the advances in agriculture and reducing waste whilst addressing shifting diets, enabled a doubling in agricultural production and a reduction in environmental impacts. In this light, recent studies have argued that current food production practices do not suffice and therefore a transformation of the food system is required (Godfray et al., 2010). Possible approaches to enhance future food availability include reduction in food waste. Food loss is the decrease in edible food mass, which occurs at production, postharvest and processing stages in the food supply chain. Recovering food loss is a huge opportunity to reduce production demand and food wastage in developed countries. A dietary shift is another possible approach to protect biodiversity in a more sustainable way. The changing dietary patterns and demand for animal products has shifted the consumption pattern, putting stress on the need to develop more sustainable sources of protein from non-animal sources to improve consumer appeal and sensor quality for different cultures (Davis et al., 2016). Improvement in productivity critically depends on development of new technologies (new crop varieties, precision agriculture). Moving towards first generation biofuels (corn sugarcane) and second-generation biofuels (cellulose material) can alleviate the food insecurity and may offer new high yielding sources of biofuel (Chen et al., 2017). Closing yield gaps in existing crop and livestock production by using new genetically tailored management technologies can increase genetic potential and improve tolerance towards different biotic and abiotic stresses. Apart from these expanding land resources for unlocking new arable land and water resources for efficient use of irrigation water which can be achieved by hydrological modelling. Development of a new farming system can intensify the sustainable use of land and water leading to higher profits in global crop production. This includes techniques to reduce soil evaporation, surface run off, soil infiltration and efficiency of irrigation systems.

## 7.6 CONCLUSION

The global food security problem is complicated because it needs a continual emphasis on both human and environmental health. All kinds of malnutrition, including food waste and loss, are predicted by food insecurity. The current food demand emphasizes the necessity to address the difficulties associated with food insecurity. Abiotic stressors, population increase, climate change, and economic growth all pose severe threats to food supply, making existing food production methods insufficient to fulfil future demands. As a result, an integrated system of interventions based on transdisciplinary research and technology innovation can aid in the achievement and maintenance of a genuinely global solution to food insecurity. To achieve this aim, policies, designs, and interventions boosting nutrition-sensitive agriculture, driving economic development, and promoting food systems that emphasize safe, nutritious, adequate, and high-quality food for all must be prioritized in all four pillars of food security.

## REFERENCES

Abiad MG, and Meho LI. 2018. Food loss and food waste research in the Arab world: A systematic review. *Food Secur* 10: 311–322.

Akbari G, Sanavy SA and, Yousefzadeh S. 2007. Effect of auxin and salt stress (NaCl) on seed germination of wheat cultivars (Triticum aestivum L.). *Pak J Biol Sci* 10(15): 2557–2561.

Aroca R, Vernieri P, Irigoyen JJ, Sánchez-Dıaz M, Tognoni F, and Pardossi A. 2003. Involvement of abscisic acid in leaf and root of maize (Zea mays L.) in avoiding chilling-induced water stress. *Plant Sci* 165(3): 671–679.

Ashkavand P, Zarafshar M, Tabari M, Mirzaie J, Nikpour A, Bordbar SK, and Striker GG. 2018. Application of SiO2 nanoparticles as pretreatment alleviates the impact of drought on the physiological performance of Prunus mahaleb (Rosaceae). *Boletín de la Sociedad Argentina de Botánica* 53(2): 207–219.

Asian Development Bank, 2013. Food Security in Asia and the Pacific. Philippines: Asian Development Bank.

Augustin MA, Riley M, Stockmann R, Bennett L, Kahl A, Lockett T, Osmond M, Sanguansri P, Stonehouse W, Zajac I, and Cobiac L. 2016. Role of food processing in food and nutrition security. *Trends Food Sci Technol* 56: 115–125.

Beddington J. 2010. Food security: contributions from science to a new and greener revolution. *Philosl Trans Soc Biol* 365(1537): 61–71.

Bockheim JG, and Gennadiyev AN. 2000. The role of soil-forming processes in the definition of taxa in Soil Taxonomy and the World Soil Reference Base. *Geoderma* 95(1–2): 53–72.

Boyer JS. 1982. Plant productivity and environment. *Sci* 218(4571): 443–448.

Carillo P, Annunziata MG, Pontecorvo G, Fuggi, A, and Woodrow P. 2011. Salinity stress and salt tolerance. *Abiotic stress in plants-mechanisms and adaptations* 1: 21–38.

Chen M, and Smith PM. 2017. The US cellulosic biofuels industry: Expert views on commercialization drivers and barriers. *Biomass and Bioenerg* 102: 52–61.

Cole MB, Augustin MA, Robertson MJ, and Manners JM. 2018. The science of food security. *NPJ Sci Food* 2(1): 1–8.

Davis KF, Gephart JA, Emery KA, Leach AM, Galloway JN, and D'Odorico P. 2016. Meeting future food demand with current agricultural resources. *Global Environ Change* 39: 125–132.

Devaux A, Kromann P, and Ortiz O. 2014. Potatoes for sustainable global food security. *Potato Res* 57(3): 185–199.

ERS U. 2017. International Food Security Assessment, 2017–2027.

FAO 2011. Global food losses and food waste – Extent, causes and prevention.

FAO I, and WFP 2015. The State of Food Insecurity in the World 2015. Meeting the 2015 international hunger targets: taking stock of uneven progress. Rome.

FAO, WFP, and IFAD 2012. The State of Food Insecurity in the World 2012. Economic growth is necessary but not sufficient to accelerate reduction of hunger and malnutrition, FAO, Rome.

Farooq M, Aziz T, Wahid A, Lee DJ, and Siddique KH. 2009. Chilling tolerance in maize: agronomic and physiological approaches. *Crop and Pasture Sci* 60(6): 501–516.

Flora SJS, Mittal M, and Mehta A. 2008. Heavy metal induced oxidative stress & its possible reversal by chelation therapy. *Indian J Med Res* 128(4): 501.

Flowers TJ. 2004. Improving crop salt tolerance. *J Exp Bot* 55(396): 307–319.

Foresight UK. 2011. The future of food and farming. *Final Project Report, London, The Government Office for Science*.

Gibson M. 2012. Food security – a commentary: what is it and why is it so complicated? *Foods* 1: 18–27.

Global Network Against Food Crises, 2021. Global Report on food Crises.

Godfray HCJ, Beddington JR, Crute IR, Haddad L, Lawrence D, Muir JF, and Toulmin C. 2010. Food security: the challenge of feeding 9 billion people. *Sci* 327(5967): 812–818.

Harris D, Tripathi RS, and Joshi A. 2002. On-farm seed priming to improve crop establishment and yield in dry direct-seeded rice. *Direct seeding: Research Strategies and Opportunities, International Research Institute, Manila, Philippines*, 231–240.

Kaya MDO, kçu G, Atak M, Cıkıll Y, and Kolsarıcı Ö. 2006. Seed treatments to overcome salt and drought stress during germination in sunflower (Helianthus annuus L.). *Europ J Agron* 24(4): 291–295.

Knight A, and Davis C. 2007. What a waste! Surplus fresh foods research project. Available at: www.veoliatrust.org/docs/Surplus_Food_Re search.pdf. Access 7/12/2010.

Leandro A, Pacheco D, Cotas J, Marques JC, Pereira L, and Gonçalves AM. 2020. Seaweed's bioactive candidate compounds to food industry and global food security. Life 10: 140.

Lewis KD, and Burton-Freeman BM. 2010. The role of innovation and technology in meeting individual nutritional needs. *J Nutr* 140: 426S–436S.

Lobell DB, and Gourdji, SM. 2012. The influence of climate change on global crop productivity. *Plant physiol* 160(4): 1686–1697.

Lundqvist J, De Fraiture C, and Molden D. 2008. *Saving water: from field to fork: curbing losses and wastage in the food chain.* Stockholm International Water Institute.

Manickavelu A, Nadarajan N, Ganesh SK, Gnanamalar RP, and Chandra Babu R. 2006. Drought tolerance in rice: morphological and molecular genetic consideration. *Plant Growth Regul* 50(2): 121–138.

Martindale, W. 2017. The potential of food preservation to reduce food waste. *Proc Nutr Soc* 76: 28–33.

McCarthy U, Uysal I, Badia -Melis R, Mercier S, O' Donnell C, and Ktenioudaki A. 2018. Global food security–Issues, challenges and technological solutions. *Trends Food Sci Technol* 77: 11–20.

McCarthy U, Uysal I, Laniel M, Corkery G, Butler F, McDonnell KP, and Ward S. 2013. Sustainable global food supply networks. In: Norton T, Tiwari B, and Holden N. (eds.) *Sustainable Food Processing* 521–538. Wiley.

Nahar K, Hasanuzzaman M, and Majumder RR. 2009. Effect of low temperature stress in transplanted aman rice varieties mediated by different transplanting dates. *J Plant Sci* 2(3): 132–138.

Okçu G, Kaya MD, and Atak M. 2005. Effects of salt and drought stresses on germination and seedling growth of pea (Pisum sativum L.). *Turk J Agric For* 29(4): 237–242.

Qadir M, Quillérou E, Nangia V, Murtaza G, Singh M, Thomas RJ, Noble AD. 2014. Economics of salt-induced land degradation and restoration. *Nat Resour* 34(8): 282–295.

Sanghera GS, Wani SH, Hussain W, and Singh NB. 2011. Engineering cold stress tolerance in crop plants. *Curr Genomics* 12(1): 30.

Santeramo FG, and Lamonaca E. 2021. Food loss–food waste–food security: a new research agenda. *Sustainability* 13: 4642.

Santeramo FG, and Shabnam N. 2015. The income-elasticity of calories, macro-and micronutrients: What is the literature telling us? *Food Res Int* 76: 932–937.

Schutzendubel A, and Polle A. 2002. Plant responses to abiotic stresses: heavy metal-induced oxidative stress and protection by mycorrhization. *J Exp Bot* 53(372): 1351–1365.

Shabnam N, Santeramo FG, Asghar Z, and Seccia A. 2016. The impact of food price crises on the demand for nutrients in Pakistan. *J South Asian Dev* 11: 305–327.

Shah F, Huang J, Cui K, Nie L, Shah T, Chen C, and Wang K. 2011. Impact of high-temperature stress on rice plant and its traits related to tolerance. *J of Agric Sci* 149(5): 545–556.

Stephens EC, Jones AD, and Parsons D. 2017. Agricultural systems research and global food security in the 21st century: An overview and roadmap for future opportunities. Agricultural Systems.

Thakur P, Kumar S, Malik JA, Berger JD, and Nayyar H. 2010. Cold stress effects on reproductive development in grain crops: An overview. *Environ Exp Bot* 67(3): 429–443.

Tripathy JN, Zhang J, Robin S, Nguyen TT, and Nguyen HT. 2000. QTLs for cell-membrane stability mapped in rice (Oryza sativa L.) under drought stress. *Theor Appli Genet* 100(8):1197–1202.

United States Department of Agriculture, 2016, Census of Agriculture.

Vollenweider P, and Günthardt-Goerg MS. 2005. Diagnosis of abiotic and biotic stress factors using the visible symptoms in foliage. *Environm Pol* 137(3):455–465.

Watanabe T, and Kume T. 2009. A general adaptation strategy for climate change impacts on paddy cultivation: special reference to the Japanese context. *Paddy Water Environ* 7(4): 313–320.

Wheeler T, and Von Braun J. 2013. Climate change impacts on global food security. *Science* 341(6145): 508–513.

World Bank. 2010. GDP per capita, PPP. International Comparison Program database.

World economic forum, 2017. Report.

World Food Program, 2021. Global report on food crises.

Wuana RA, and Okieimen FE. 2011. Heavy metals in contaminated soils: a review of sources, chemistry, risks and best available strategies for remediation. *Int Sch Res Notices* 2011: 1–21.

Xin Z, and Browse J. 2000. Cold comfort farm: the acclimation of plants to freezing temperatures. *Plant Cell Environ* 23: 893–902.

Zeid IM, and Shedeed ZA. 2006. Response of alfalfa to putrescine treatment under drought stress. *Biol Plant* 50(4): 635–640.

# CHAPTER 8

# Food Safety, Quality, and Policies

**Ajay Rathore[1]* and Ambika Bhatia[2]**
[1]Punjabi University, Patiala
[2]Punjabi University Regional Centre for Information Technology & Management, Mohali
*Corresponding author email: ajay_0172@yahoo.co.in

## CONTENTS

- 8.1 Introduction .......... 110
- 8.2 Food Quality .......... 111
- 8.3 Food Safety and Standards Authority of India (FSSAI) .......... 112
- 8.4 Food Safety .......... 112
    - 8.4.1 Background .......... 113
    - 8.4.2 Types of Food Safety Risks .......... 114
        - 8.4.2.1 Food Infection/Food Poisoning .......... 115
        - 8.4.2.2 Food Intoxication .......... 115
    - 8.4.3 Challenges to Improving Food Safety in India .......... 115
- 8.5 Food Policies in India .......... 116
    - 8.5.1 Food Safety Management Systems (FSMS) .......... 116
    - 8.5.2 Why Should You Use HACCP? .......... 117
    - 8.5.3 Need for the FSSA Act .......... 117
    - 8.5.4 Regulations of FSSA .......... 117
        - 8.5.4.1 Labeling and Packaging .......... 117
        - 8.5.4.2 Signs and Customer Notifications .......... 118
        - 8.5.4.3 Registration, Licensing, and Health and Sanitary Permits .......... 119
    - 8.5.5 Penalties .......... 119
- 8.6 Conclusion .......... 119
- References .......... 120

## 8.1 INTRODUCTION

Food is a key factor of population health, nutritional condition, and productivity. We often read in newspapers or hear news everyday about issues related to health that are caused because of adulterated food. As a result, it is critical that the food we eat be nutritious and safe. Unsafe food can cause a variety of foodborne illnesses. Foodborne diseases and fatalities cost the United States $77 billion per year (Scharff, 2012). The adulteration of food items at such facilities has played a key role in recent food safety issues, raising worries. As per National Family Health Survey, more than 2 lakh children under the age of five years old were affected with acute diarrhea in India during the period of 2015–2016. Food-borne disease may cause death as well as harm to commerce and tourism, as well as loss of wages, create unemployment, and lawsuits, all of which can stymie economic progress. As a result, food safety and quality have taken on a global relevance. The Covid 19 pandemic is a classic example for the same. Initial media reports claimed that the Covid 19 virus spread from China's meat market, which eventually impacted not only China, but entire world's population.

Food systems in many developing nations are undergoing significant changes as a result of globalization, urbanization, and growing living standards. Food safety and quality are vital at home, but they are especially important in large-scale food manufacturing and processing, as well as where food is prepared and served fresh. In years gone by, many foods were prepared in small set ups or at home. A range of processed foods, food for health / functional foods have been created as a result of advancements in technology and processing, higher per capita incomes and improved purchasing power, as well as increasing customer demand. The safety of such foods must be determined. Furthermore, conventional food safety techniques, such as Hazard Analysis and Critical Control Point (HACCP), are less capable of meeting the need for identifying, quantifying, controlling, and managing these food safety hazards, particularly in grain handling and processing plants (Sperber, 2005). Food quality and safety are becoming increasingly essential, posing new problems for supply chain players as they try to keep up with changing customer demand. Products having a set of special characteristics that are targeted to meet the demands of customers are necessary (Henson and Reardon, 2005). The safety concerns that have confronted the world, as well as India, have altered dramatically in the last decade, and issues connected to food quality and safety have taken on a new level of importance. Few factors which are responsible are mentioned:

- **Packed food:** There are many processed and packed food that are available in market and it is very important for customers to know if they are safe to use or not.
- **Eating habits and change in lifestyle:** There has been a drastic change eating habits and in the lifestyle of people, especially the younger generation. People are dining outside their homes in greater numbers. Food is made in bulk and handled by a large number of people in commercial settings, which increases the risk of contamination. Furthermore, food is made many hours ahead of time and may deteriorate if not properly preserved.

- **Transportation of food**: The logistics regulating bulk food transportation are complicated, and there is a lengthy time between preparation and consumption. As a result, risk analysis and safety management are important throughout mass manufacturing and distribution.
- **Pesticides**: Contamination in the environment, land, and groundwater, as well as pesticide usage in farming, all contribute pollutants. Food analysis for diverse components—both nutrients and contaminants—is also required due to the use of additives such as stabilizers, colorants, flavoring agents, and other compounds such as stabilizing agents.
- **Food-borne microbial illnesses**: Food-borne microbial illnesses are becoming more common as a result of microbiological changes, antibiotic resistance, changed human sensitivity, and worldwide travel. In the last 25–30 years, about half of all known food-borne infections have been identified. There are still a lot of food-borne diseases with unclear causes. This is a worldwide public health risk, and active monitoring networks must be established both domestically and internationally to detect, identify, and recognize new infections.

## 8.2 FOOD QUALITY

Food quality refers to the quality of food that is acceptable to customers. This includes exterior aspects such as appearance (size, shape, color, gloss, and consistency), texture, and flavor, as well as internal variables such as federal grading criteria (e.g., of eggs) (chemical, physical, microbial). Food quality is an important criterion in food manufacturing since food consumers are vulnerable to any type of contamination that may occur during the manufacturing process. Due to dietary, nutritional, or medical concerns, many customers rely on manufacturing and processing regulations to know what components are included. Due to the general increase in understanding of its potential influence on public health, food security, and trade competitiveness, developing nations are paying closer attention to food quality. Food quality is still a big concern in India's food system. The concept "food quality" refers to the characteristics that impact a product's consumer value. This covers both bad and good characteristics such as spoilage, infection, adulteration, and food safety risks, as well as positive characteristics like color, flavor, and texture. As a result, it is a holistic notion that incorporates nutritional characteristics, sensory features (color, texture, form, appearance, taste, flavor, and smell), social considerations, and safety. Safety is a prerequisite for quality and a trait that precedes it. Various countries and international organizations have established food standards that manufacturers and suppliers are obliged to follow in order to guarantee that meals are safe and of excellent quality. As a result, all food service providers (those engaged in preparation, processing, packing, and delivery) should follow standards of quality to assure food safety. The following are important considerations:

1. Raw material and water quality
2. Premises, staff, equipment, food preparation and storage, and serving spaces are all clean

3. Food storage at the proper temperature
4. Food safety
5. Excellent customer service techniques

Food hygiene may be described as being clean and not causing disease. One of the key concerns is providing healthy food to customers in increasingly competitive worldwide marketplaces. Food borne viruses and chemicals resulting in food related incidents are a key driving force for this increased focus on food safety.

## 8.3 FOOD SAFETY AND STANDARDS AUTHORITY OF INDIA (FSSAI)

It is a statutory agency created by the Government of India's Ministry of Health and Family Welfare. The FSSAI was created under the Food Safety and Standards Act of 2006, which is a consolidated legislation pertaining to food safety and regulation in India. There are several worldwide quality institutions that evaluate food items in order to show all customers which products are of higher quality. The worldwide Monde Selection quality award, founded in 1961 in Brussels, is the oldest in judging food quality. During the degustations, the goods must fulfill the Institute's selection standards, which include sensory analysis, bacteriological and chemical analysis, nutrition and health claims, and use notice. In a nutshell, the evaluations are based on the following criteria: flavor, health, convenience, labeling, packaging, environmental friendliness, and innovation.

The food processing industry, one of India's largest, is widely regarded as a "sunrise industry" in India, with enormous potential for boosting the agricultural economy, creating large-scale processed food production and food chain facilities, and generating jobs and outsource earnings as a result.

The percent of food contaminated has grown from 12.8 percent in 2011–2012 to 28 percent in 2018–2019, according to the Food Safety and Standards Authority of India (FSSAI). One out of every five foods are said to be contaminated. According to research by Food Sentry, India was the top food offender in 2013, with 11.1 percent of 3,400 confirmed food safety incidents collected from 117 nations. China came in second with 9.9%, followed by Mexico (7.5%), France (5.5%), and the United States (5.1%). Furthermore, according to FSSAI testing, 90% of food fraud in India occurs in Andhra Pradesh, Kerala, Madhya Pradesh, Maharashtra, and Uttar Pradesh. Seafood, dairy products, meats, and spices are the most commonly abused food industries. As per an article published on website Food Navigator-asia.com, it mentioned that more than a quarter of food samples in the Indian state of Haryana failed food quality tests conducted by the local Food and Drug Administration (FDA), despite the Food Safety and Standards Authority of India's (FSSAI) stating that they are improving at implementing rules.

## 8.4 FOOD SAFETY

Food safety refers to the idea that when food is consumed according to its intended usage, it will not damage anyone. The term "hygiene" refers to something that is tidy and therefore does not create sickness. One of the primary concerns is ensuring that customers have access to safe food in increasingly complex market marketplaces. Foodborne diseases and chemicals that cause food related incidents are a major driving force behind this increased focus on food safety. In most poor nations, the burden of foodborne disease caused by contamination has been overlooked. According to WHO projections, in developing countries—more than 1000 million cases of acute diarrhea in children under the age of 5 occur every year. The SARs epidemic in East Asia in 2003 is believed to have cost the territory around 2% of its GDP in the second quarter of that year, despite the fact that just 800 people died as a result of the virus. Otsuki et al. (2001) stated that many developing nations are getting a wake-up call about the difficulty of satisfying both government and commercial sanitary and phyto-sanitary (SPS) requirements in export markets as they attempt to boost agricultural exports, particularly to OECD (Organization for Economic Co-operation and Development) countries. Over the last decade, private norms or supplier norms have gained traction as a way to assure compliance with official laws, address perceived gaps in such regulations, and/or differentiate business or industry goods from those of rivals. Food safety and quality management concerns are increasingly being included in private standards. Food safety and quality management issues are increasingly being combined in private standards (e.g., the recent establishment of ISO 22000), or procedures that integrate food safety, environmental, and social (child labor, labor conditions, animal welfare) factors (Willems et al., 2005). Increasing globalization of commerce, on the other hand, raises the danger of cross-border transmission of foodborne diseases. Current disease outbreaks in the United States caused by foreign—made food produce, like cyclospora from raspberries, hepatitis A from strawberries, and salmonella from cantaloupe indicate the possible food standards obstacles that may arise in a more globalized market for developing countries.

### 8.4.1 Background

Until the green revolution in 1970, India was a food-deficit nation. The first watershed moment in Indian food regulation was the passage of the Prevention of Food Adulteration Act in 1954, followed by a related rule in 1956. The demand for on-the-go consumption has only increased as the working population has grown. At the same time, consumer knowledge of nutrition and health has grown significantly and has permeated deep into the brains of Indian consumers. On the regulatory front, the Food Safety and Standards Authority of India (FSSAI) has been raising awareness about food safety concerns across the country, as well as tightening legislation to guarantee that only safe food is made accessible to consumers, both by producers and by industry. According to Pawan Agarwal, CEO of FSSAI in an article published in Economic Times—they have adopted a more holistic strategy towards food safety instead of the previously fragmented approach.

Despite the fact that the Food Safety and Standards Act has been in effect for 11 years this year, the Food Safety and Standards Authority of India (FSSAI) was established in 2008. Since then, the organization has taken on the task of building a regulatory framework that is distinct from that of other nations yet being appropriate for India's unique characteristics.

There have been recent cases of substandard food reaching consumers, and recordings of the same have been extensively distributed on social media. As a result, food and beverage makers' primary priority is food safety. And technology is playing an important role in ensuring that product integrity is maintained throughout the value chain. Concerns about food safety are gaining traction in India. The country's rural development plan, which includes increasing agricultural exports as a method of fostering rural growth and poverty reduction, is colliding with stricter food safety and SPS requirements in potential markets. From a domestic standpoint, the 1.2 billion-strong national market is undergoing fast transformation. Increasing incomes, a burgeoning middle class, greater urbanization and literacy, and a population well-tuned to worldwide trends, all fuelled by the information advancement of technology, are producing a broad customer base that values food quality and safety more than ever. Improving food safety systems to satisfy domestic and export standards, on the other hand, is beset by governmental, regulatory, infrastructure, and institutional roadblocks.

### 8.4.2 Types of Food Safety Risks

Risks to human health from food safety are caused by a variety of factors. The potential dangers posed by these agents have an influence on the entire food supply chain, from input supply through farm to consumer table (see Figure 8.1). It may include:

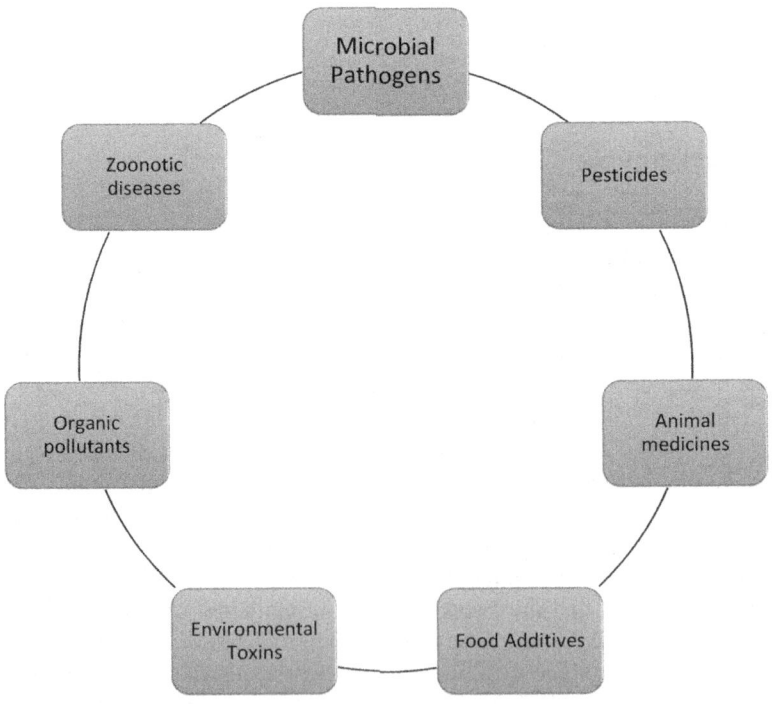

**Figure 8.1** Various causes of food safety risks.

## 8.4.2.1 Food Infection/Food Poisoning

Ingestion of live pathogenic organisms, which proliferate in the body and cause sickness. Salmonella is a good illustration of this. This bacterium can be found in the intestines of mammals. Eggs and raw milk are also good sources. Although heat kills Salmonella, insufficient cooking permits some germs to survive. Cross-contamination is a common way for Salmonella to spread. This can happen when a cook chops raw meat/poultry on a chopping board and then uses it for another meal that does not require cooking, such as salad, without cleaning it first. If an infected food operator does not clean hands with water and soap after using the restroom and before contacting food, food might get infected with Salmonella. If an infected food operator does not clean hands with water and soap after using the restroom and before contacting food, food might get infected with Salmonella.

Salmonella may multiply rapidly, doubling in number every 20 minutes. Salmonella infection symptoms include diarrhea, fever, and stomach cramps.

## 8.4.2.2 Food Intoxication

Even after a pathogen has been eliminated, certain bacteria generate dangerous toxins that can be found in food. Toxins are produced by organisms when food is not heated or chilled enough. Toxins in food are undetectable by smell, sight, or taste. As a result, meals that smell and look wonderful are not always safe. Staphylococcus aureus is an example of such an organism. These creatures can be found in the air, dust, and water. They are also found in the nasal tube, throat, and on the skin and hair of 50% of healthy people. People who have this organism contaminate food by touching these areas on their bodies while handling food. One of the indications of this pollution is diarrhea.

Human hosts, animal hosts, and their interactions with people, the virus itself, and the environment, including how food is produced, processed, handled, and stored, are all key factors in pathogen emergence. Changes in host susceptibility owing to hunger, age, and other factors, for example, might allow novel diseases to develop in susceptible populations. New strains with the potential to cause illness can be created via genetic exchange or mutations in the organisms. Pathogens can arise in new populations or geographical regions as a result of changes in eating habits, climate, mass manufacturing, food processing, and greater globalization of the food supply.

### 8.4.3 Challenges to Improving Food Safety in India

A multitude of structural, regulatory, institutional, technological, and cultural hurdles are preventing India from improving food safety, whether for the home market or for export.

- Inadequate grades and standards for the domestic market and poor enforcement
- Lack of pro-activity in addressing food-safety issues.
- Lack of good agricultural, manufacturing and hygiene practices.

- Poor services in marketing system
- Poor Infrastructure
- Weak public agriculture extension systems

## 8.5 FOOD POLICIES IN INDIA

The food industry, one of India's major businesses, is commonly regarded as a "sunrise industry" with enormous potential for boosting the agricultural economy, establishing large-scale processed food production and food chain facilities, and generating employment and export profits. The Indian food processing sector is governed by many regulations that control sanitation, licensing, and other licenses required to start and manage a food company. The Prevention of Food Adulteration Act, 1954 was India's food safety legislation (hereinafter referred to as "PFA"). The PFA had been in existence for almost five decades, and there was a need for reform for a variety of reasons, including changing food sector requirements.

### 8.5.1 Food Safety Management Systems (FSMS)

Food safety and quality concerns have evolved throughout time to include more than only the prevention of food-borne viruses, chemical toxicants, and other dangers. A food hazard can enter/come into the food at any point along the food chain, thus proper management is required throughout the chain. The following measures can be taken to assure food safety and quality:

- Hazard Analysis Critical Control Points (HACCP)
- Good Manufacturing Practices (GMP)
- Good Handling Practices (GHP)
- **GMP**: are a type of quality assurances that ensure that manufacturers and processors take proactive efforts to guarantee that their goods are safe. It allows for the reduction or elimination of contamination and misleading labeling, saving consumers from being misled and assisting in the purchase of safe items. GMP is a useful business tool for improving compliance and performance among manufacturers and producers.
- **GHP**: Good Handling Practices provide a complete strategy to identifying possible sources of risk from the farm to the shop or consumer, as well as the processes and procedures used to reduce the risk of contamination. It guarantees that all those who handle food do it in a hygienic manner.
- **HACCP**: It stands for Hazard Analysis and Critical Control Points, and it is a method of ensuring food safety. HACCP is a method of food manufacturing and storage in which raw materials and each individual step in a process are scrutinized for their potential to contribute to the growth of pathogenic microorganisms or other food risks. It entails identifying risks, assessing the likelihood of hazards occurring at each step/stage in the food chain—raw material acquisition, production, distribution, and food product use—and establishing strategies to mitigate hazards. HACCP has become a requirement for both export and food supply to large retailers (E. Taylor, 2008; J. Z. Taylor, 2008).

The Food Safety and Standards Act, 2006 (hereafter referred to as "FSSA"), which took effect in place of the PFA, supersedes all existing food-related laws. It explicitly abolished eight statutes that were in effect previous to the implementation of FSSA:

1. The Prevention of Food Adulteration Act, 1954
2. The Fruit Products Order, 1955
3. The Meat Food Products Order, 1973
4. The Vegetable Oil Products (Control) Order, 1947
5. The Edible Oils Packaging (Regulation) Order, 1998
6. The Solvent Extracted Oil, De oiled Meal, and Edible Flour (Control) Order, 1967
7. The Milk and Milk Products Order, 1992
8. Essential Commodities Act, 1955 (in relation to food)

### 8.5.2 Why Should You Use HACCP?

1. It's a proactive strategy for ensuring food safety.
2. It ensures high-quality production on a regular basis.
3. Although necessary, end-of-product inspection and testing is time consuming, costly, and only discovers issues after they exist. HACCP, on the other hand, enables us to detect risks at any step of processing or manufacturing in order to assure a high-quality end product by taking necessary action at the point of occurrence.
4. It allows producers, processors, distributors, and exporters to make efficient and cost-effective use of resources to ensure food safety.
5. Through HACCP, GMP, and GHP, the FSSA of 2006 assigns main responsibility for food safety to manufacturers and suppliers.

### 8.5.3 Need for the FSSA Act

The FSSA begins the process of harmonizing India's food rules with international norms. It creates a new national regulatory body, the Food Safety and Standards Authority of India (hereinafter "FSSAI"), to develop science-based food standards and to regulate and monitor food manufacturing, processing, storage, distribution, sale, and import in order to ensure the availability of safe and wholesome food for human consumption. All food imports will thus be subject to the requirements of the FSSA as well as the rules and regulations that were announced by the government on August 5, 2011.

### 8.5.4 Regulations of FSSA

#### 8.5.4.1 Labeling and Packaging

The Food Safety and Standards (Packaging and Labeling) Rules, 2011 (hence referred to as the "Packaging and Labeling Regulations") establish distinct packaging and labeling regulations that spell out the legislative and regulatory requirements for product packaging and labeling. A simple perusal of the Packaging and Labeling Laws reveals that there are several types of items: pre-packaged, proprietary, and other products specifically listed in the regulations. "Proprietary food," as defined by Regulation 2.12 of the Food Safety and Standards (Food Products Standards and Food Additives) Regulations, 2011, is food which has not been defined under these regulatory frameworks. "Prepackaged" or "pre-packed food" is defined in Regulation 1 (8) of the Packaging and Labeling Regulations as food that has been placed in a packaging of any kind in such a way that the contents cannot be altered without tampering and is ready for sale to the customer.

The Packaging and Labeling Regulations provide the following general criteria for food product labeling under the FSSA:

- The declaration particulars that must be included on the label under these Regulations must be written in English or Hindi in Devnagri script: Provided, however, that nothing in this rule prohibits the use of any other language in addition to the one specified by this regulation;
- Pre-packaged food may not be characterized or displayed on any label or in any way that is untrue, misleading, or deceptive, or that is likely to give the wrong impression about its character in any way;
- When a wrapper covers the container, the wrapper must convey the essential information, or the container's label must be easily visible through the wrapper and not hidden by it.

Because a wide range of food products are starting to be imported into India, the Packaging and Labeling Regulations require that the country of origin of the food be mentioned on the label of food imported into India, and that when a food undergoes processing in a second country that changes its nature, the country in which the processing is performed shall be considered the country of origin of the food.

As a result, the above are the statutory and regulatory criteria that must be followed when labeling products sold as "pre-packaged goods" in the Indian market.

### 8.5.4.2 *Signs and Customer Notifications*

After briefly discussing the statutory and regulatory requirements for product labeling, it is important to have a better understanding of the statutory and regulatory requirements for signage and customer notices from the perspective of a food establishment. It is essential to note that, while the FSSA does not expressly provide for any legislative or regulatory obligations for signs or customer notifications, it does provide provisions for food company owners to market their products.

The term "advertisement" (which includes a "notice") is defined in Section 3 (1) (b) of the FSSA as any audio or visual publicity, depiction, or proclamation made by means of any light, audio, fumes, gas, publish, digital media, internet, or webpage, and includes any notification, circular, branding, wrappings, receipt, or other documents.

Section 24 of the FSSA states that no food must be advertised in a way that is misleading or deceptive, or that violates the laws, rules, and regulations enacted thereunder. No one shall engage in any unfair trade activity for the purpose of promoting the sale, supply, use, or consumption of food items, or engage in any unfair or deceptive conduct, including the practice of making any statement, whether verbally, in writing, or by visible depiction, which:

- Fraudulently advertises that the foods are of a specific standard, quality, quantity, or grade-composition;
- Falsely or misleadingly promotes the necessity for, or usefulness of;
- Provides to the public any assurance of effectiveness that is not based on sufficient or scientific explanation, provided that if a defense is offered to the effect that such guarantee is based on adequate or scientific rationale, the burden of proof shall fall on the person making such defense.

Because the FSSA applies to all food industry operators in India, the clause regarding ads must be followed.

### 8.5.4.3 Registration, Licensing, and Health and Sanitary Permits

It is also worth noting that FSSA, being the sole regulation relevant to the food sector throughout the country, will also apply to national health and sanitary licenses.

According to Regulation 2.1 of the License and Registration Regulations, all food manufacturing operators in the nation must be enrolled or licensed in full compliance with the License and Registration Regulations; as a result, no person may start a food company or business unless he or she has a valid license and meets the safety, sanitation facilities, and good hygiene requirements. The food company is held responsible for adhering to the labeling, safety, and health and sanitary standards outlined in the License and Registration Regulations. The labeling standards are outlined in the rules and must be followed at all times, particularly in the case of pre-packaged items.

### 8.5.5 Penalties

The FSSA imposes penalties in the event of violation. Non-compliance with the FSSA's different requirements might result in a penalty of up to 2 lakh rupees. However, in Section 63, any individual or food enterprise operator (except those exempted from licensing under sub-section (2) of Section 31 of FSSA) who manufactures, tries to sell, store, distribute, or import any article of food without a permit, personally or by any person on his behalf who is required to obtain a license, shall be punished with imprisonment for a term which may extend to six months and also with a fine which may extend to Five Lakh Rupees.

## 8.6 CONCLUSION

Development of transportable, quick, inexpensive, and highly sensitive analytical devices is needed for analysis of food safety and quality control to fulfil the demands of consumers. Good quality food delivers basic human needs as it fulfils nutrition security, improves national budget, hospitality sector, etc. in order to promote sustainable development. Consumers' demands for different types of food have been increased due to increase in the global population which has resulted in complex and longer global food chain. Both benefits and consequences for food safety occurs due to the intensification and industrialization of agriculture. Every individual should perform their duty to ensure food safety.

## REFERENCES

Henson S and Reardon T. 2005. Private agri-food standards: Implications for food policy and the agri-food system. *Food Policy* 30(3): 241–253.

Otsuki T, Wilson JS, and Sewadeh M. 2001. Saving two in a billion: quantifying the trade effect of European food safety standards on African exports. *Food Policy* 26(5): 495–514.

Scharff RL. 2012. Economic burden from health losses due to foodborne illness in the United States. *J Food Prot* 75(1): 123–131.

Sperber WH. 2005. HACCP does not work from farm to table. *Food Control* 16(6): 511–514.

Taylor E. 2008. HACCP for the hospitality industry: history in the making. *International Journal of Contemporary Hospitality Management* 20(5): 480–493.

Willems S, Roth E, and Van Roekel J. 2005. Changing European Public and Private Food Safety and Quality Requirements. Agriculture and Rural Development Discussion Paper, 15.

CHAPTER 9

# Emerging Role of Nanotechnology in the Food Industry for Food Processing and Packaging

**Nitika Kapoor**[*]
PG Department of Botany, Hans Raj Mahila Maha Vidyalaya, Jalandhar (Punjab), India
*Corresponding author Email: nitikaarora8@gmail.com

## CONTENTS

| | | |
|---|---|---|
| 9.1 | Introduction | 122 |
| 9.2 | Role of Nanotechnology in Food Processing | 123 |
| | 9.2.1 Nanoencapsulation and Nano-Carriers: Protect Aroma, Flavor and Other Ingredients in Food | 124 |
| | 9.2.2 Anticaking and Gelling Agents: Consistency and Texture Modifiers | 126 |
| | 9.2.3 Nanoadditives and Nutraceuticals: Improve Nutritional Value of Food | 127 |
| 9.3 | Role of Nanotechnology in Food Packaging | 127 |
| | 9.3.1 Bio-Based Packaging | 128 |
| | 9.3.2 Improved Packaging | 130 |
| | 9.3.3 Active Packaging | 131 |
| |     9.3.3.1 Antimicrobial Active Packaging/Antimicrobial Films | 132 |
| |     9.3.3.2 Oxygen Scavenging Films | 133 |
| |     9.3.3.3 UV Absorbing Films | 133 |
| | 9.3.4 Smart/Intelligent Packaging | 134 |
| |     9.3.4.1 Bionanosensors | 134 |
| |     9.3.4.2 Oxygen Indicators | 135 |
| 9.4 | Nanotechnology and Food Safety Issues | 135 |
| 9.5 | Conclusion | 136 |
| References | | 137 |

DOI: 10.1201/9781003258568-9

## 9.1 INTRODUCTION

In recent years nanotechnology has become increasingly important in the area of food sciences. Novel nanomaterials developed during current years used frequently to improve food quality, safety, crop growth and to monitor environmental conditions that bring new revolution in the food industry (Primožič et al., 2021). Nanotechnology is the technology that deals with the atoms, molecules, or the macromolecules on the nanoscale to create and use materials that have novel properties as compared to their macroscale counterparts. The naomaterials of the range 1 to 100 nm, have novel properties due their high surface to volume ratio and other unique physiochemical properties like color, solubility, strength, diffusivity, toxicity, optical and magnetic etc. (Gupta et al., 2016). The rising consumer awareness about food quality and its health benefits driving the researchers to develop such technologies that can enhance food quality without disturbing the nutritional value of the product. So under these circumstances, the demand of nanoparticles has increased in the food industry, which plays an important role in processing and packaging of food products that not only increased the shelf life of food products but also enhances their nutritive status. Nanoparticles are frequently utilized in the food industry to mask unwanted flavors, improve the texture and consistency of food and enhanced nutraceutical value of food products as they are found to be non-toxic and stable at high temperature and pressures (Sawai, 2003; Singh et al., 2017).

The application of nanotechnology in the food industry can be categorized under two major categories *viz.* nanostructured food ingredients and food nanosensing. Nanostructured food ingredients covered a wide area from food processing to food packaging. In food processing, nanomaterial can be used as anticaking/gelling agents that protect aroma and flavor of food, increase bio-availability of bioactive compounds; and as nanoadditives (nanocomposite, nanoemulsifiers) that improve nutritional value of food. Nanotechnology has also uplifted the area of food packaging with modern and efficient methods of packing that provide microbial free, nutritious and fresh food products to consumers with environment friendly biodegradable packaging material. Nanotechnology has the potential to improve health, wealth and quality of life, as well as reducing impact on the environment. About more than 500 nanopackaging materials are in commercial use, while the further use of nanotechnology is predicted to be enhanced in the food packaging industry within the next decade (Dasgupta et al., 2015). Nanoparticles used in nanopackaging are divided into two categories: nano-object materials (nanoscale dimensions less than 100 nm) and nano-structured materials. In nano-objects, nanomaterials are used as filler that involves the use of nanoplate, nanoparticles and nanofibers. While in nano-structured materials, the nanomaterials are dispersed into a polymer matrix to form nanocomposites, material with nanostructured surfaces, assembled nanomaterial and shell structure etc. (Kuswandi, 2016; Pathakoti et al., 2017).

Further nano-packaging can be designed to release antimicrobials, antioxidants, enzymes, flavours and nutraceuticals to extend shelf life of the products. Various nanomaterial based anti-microbial films are available in a market that improves the shelf life of food and dairy products. Other important aspects of nanotechnology are the development of nanocomposites that could help to reduce the waste of the packaging associated with processed foods. The use of nanocomposites based edible and biodegradable films in packaging will enhance the shelf life of product along with reduction in packaging waste. These types of packaging material will also be environmentally friendly as they reduce the burden of non-biodegradable packaging material on environment and also on stray animals (Lagaron et al. 2005; Pathakoti et al., 2017). On the other hand food nanosensing proposed promising domain of the bionanosensors that play an important role in providing accurate information to consumers regarding quality and shelf life of food products (Dasgupta et al., 2015; Baliyan et al., 2020). In the present chapter, the role of nanotechnology is discussed under two major headings i.e. food processing and food packaging along with safety issues associated with nanoprocessed foods.

## 9.2 ROLE OF NANOTECHNOLOGY IN FOOD PROCESSING

Nanotechnology plays an important role in today's food processing industry as the nanostructured food ingredients offer better taste, texture, and consistency (Singh et al., 2017). Nanotechnology also helps to increase the shelf life of different types of food materials by bringing down the rate of microbial infestation and by carrying the food additives to different parts of food without disturbing their original morphology (Pradhan et al., 2015). Size of nanoparticles also regulates the delivery of different bioactive compounds to various target sites of body as different cell line absorbs differently sized nanoparticles more efficiently than the others (Ezhilarasi et al., 2013). An ideal nanoparticle delivery system should possess the precise qualities *viz.* ability to deliver the active compound precisely at the target location, should be available at a target time and at specific rate, should able to maintain appropriate concentration of active compounds for a longer period of time under storage conditions.

With these qualities nanotechnology is being used to form biopolymer matrices, emulsions, encapsulation, simple solutions, and association colloids which offer efficient delivery systems. Nano-encapsulation is much better than traditional methods of encapsulation as nanoencapsulation of active food particles conceals odor/taste, control interactions and release of active ingredients, penetrate deeply into tissues due to their smaller size for efficient delivery of active compounds at target site in the body, ensuring availability at a target time and in specific concentration, protecting them from physical or biological degradation during processing, storage, and utilization, and also exhibiting compatibility with other compounds in the system (Ubbink and Kruger, 2006, Weiss et al., 2006; Singh et al., 2017). The key roles of nanotechnology in food processing industry including nanoencapsulation, anticaking/gelling agents, nanoadditives and nutraceuticals are summarized in Figure 9.1.

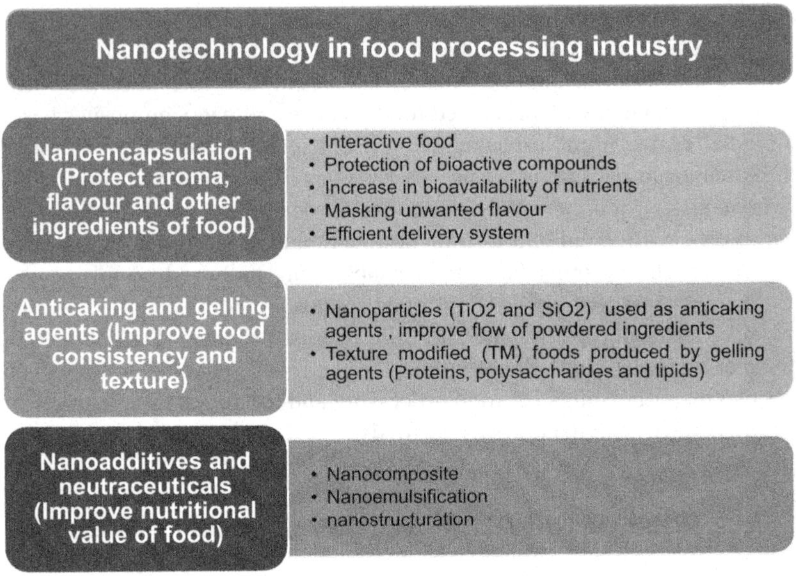

**Figure 9.1** Application of nanotechnology in food processing.

### 9.2.1 Nanoencapsulation and Nano-Carriers: Protect Aroma, Flavor and Other Ingredients in Food

To improve the taste, texture, flavor, colour, aroma and bioavailability of nutraceuticals and health supplements, food nanotechnology industry uses various latest techniques including nanoencapsulation and nano-carriers. Nanoencapsulation is one of the most practical techniques that protects and allows for controlled release of bioactive compounds at a particular time and place. Nutrients that are less soluble in water like vitamins and antioxidants are targeted for specific site by nanoencapsulation. In the present scenario, nanoencapsulation technology has great potential to be used in food and agriculture industries. Nanoencapsulation involves entrapping of bioactive agents within carrier materials with a dimension in nano-scale to form nano capsules which are within the size range of 10 to 1000 nm (Baliyan et al., 2020, Sekhon, 2010). Massive demands of functional food with higher nutritional value and lower dose of synthetic preservatives lead to numerous applications of nanoencapsulation in food processing.

This technology has enhanced the stability of sensitive compounds during their production, storage and ingestion. For example, nanoencapsulation reduces the evaporation and degradation of important vitamins, also masks the unpleasant taste by providing pleasant aromas and also limits exposure to water oxygen and light that otherwise temper with quality of products (Fathi et al., 2014).Encapsulation can also modify the physical characteristics of original material which further helps in easy handling, separating the components of mixture which otherwise can react and also help in providing accurate concentration and uniform distribution of active agents (Desai and Park, 2005).

Nanoencapsulation technique in the food processing industry is mainly focused on food preservation and interactive foods. To deliver nutrients and additives, developing new tastes, to improve the texture of food, to increase the bioavailability of nutritional components and to increase shelf life, nanoparticles are added into existing food (Abbas et al., 2009). Interactive foods are one of the important outcomes of nanoencapsulation technique. Interactive foods are modified as per the requirement and taste of consumers and provide many possibilities like flavor and taste masking, controlled release, better dispersibility of water-insoluble food ingredients and additives, change in color and flavor, warming/cooling effects, long lasting flavor and foam producing characteristics etc (Ludwig, 2006). One of the common examples of the nanoencapsulation is Mystery Colorz Cheetos®, a food product that dyed the tongue green or blue when cheese puffs come in contact with the saliva in the mouth. In beverages and instant refreshing drinks, gas infusing nanoparticles and flavor delivery system are used to develop froth or flavor (Norris et al., 2011). Encapsulation technology also helps to protect bioactive compounds present in food items by providing a physical barrier, by increasing their stability during processing and storage and slowing down the degradation processes like oxidation/hydrolysis until the product is delivered at the target sites (Fang and Bhandari, 2010; de Vos et al., 2010).

As compared to micro-size carriers, nanocarriers provide more surface area and thus improve solubility of water insoluble food additives, increase bioavailability and expedite controlled release and targeting of the encapsulated food ingredients (Mozafari, 2005; Weiss et al., 2006). "Delayed release" and "sustained release" are two mechanisms which ensure the delivery of bioactive compounds as per their requirement in any system. In case of delayed release mechanism, the release of bioactive compounds delayed from a restricted "lag time" up to a point when its release is desired. Flavor/color release in ready-meals and in beverages and protection of certain nutritious compounds in gastric condition followed by their release in intestine use "delayed release mechanism". Whereas "sustained release mechanism" maintain constant concentration of bioactive compounds at its target site and extends the release of the encapsulated material, including flavor or certain drugs. In one of the studies carried out by Santos et al. (2013) it was observed that nanoparticles as carriers enhanced the aqueous solubility, stability, bioavailability and target specificity of quercetin, epigallocatechin gallate, curcumin and resveratrol.

Furthemore, the methods used for encapsulation technology are broadly categorized into two methods *viz.* Physical methods: Spray drying, Spray cooling/chilling, Spinning/rotating disc, Fluidized bed drying, Extrusion, Coextrusion and Lyophilization; Chemical methods: Coacervation, Liposome and Inclusion complexation. Regarding encapsulating material used, they should be of food grade, biodegradable, and stable in food systems during its processing, storage, and consumption. In addition to these properties, encapsulating material should have outstanding barrier properties, film-forming properties and phase transition points at which the matrix undergoes phase transformation. The majority of encapsulating material used in the food industry are based on lipids (fatty acids, wax, phospholipids, glycerides), carbohydrates (cellulose derivatives, starch derivatives, plant extract) or protein (vegetable/animal proteins) (Gaonkar et al., 2014; Baliyan et al., 2020). Nanoencapsulation techniques help to improve the flavor release and retention and to improve culinary balance (Nakagawa, 2014). Rutin is a common dietary flavonoid with essential pharmacological properties, but its use in the food business is limited due to its poor solubility. When compared to free rutin, the ferritin nanocages encapsulation improved the solubility, thermal, and UV radiation stability of ferritin trapped rutin (Yang et al., 2015). Nanoemulsions are increasingly being used to distribute lipidsoluble bioactive chemicals since they can be made using natural food ingredients and can be tailored to improve water dispersion and bioavailability (Ozturk et al., 2015). Encapsulation with nanoparticles also enhanced the extent of bioavailability of drugs and various other nutraceutical compounds (Dekkers et al., 2011).

### 9.2.2 Anticaking and Gelling Agents: Consistency and Texture Modifiers

Anticaking agents are additives placed in powdered or granulated materials, to prevent the formation of lumps (caking), which further help in packaging, transport, flowability, and consumption of food products. Caking mechanisms depend on the nature of the material. Titanium dioxide and silicon dioxide ($SiO_2$) nanomaterials are commonly used as color or flow agents in food items that also enhance fragrances or flavors of food products (Dekkers et al., 2011). Silica ($SiO_2$) is commonly used anti-caking agent that improves flow property of powered ingredients and also act as carrier for flavors or active compounds in food. $SiO_2$ nanoparticle (NP) is registered as a food additive -E551 in the European Union (EU) and used as an anti-caking or anti-clumping agent (Wang et al., 2013).

Gelling agents are another class of food additives that help to improve texture of food. They are used to thicken and stabilize various foods, like jellies, desserts and candies etc. These agents provide the foods with texture through formation of a gel. Nowadays demand for healthy and soft food with modified texture has been increasing to provide nutritious food to elderly population (Aguilera & Park, 2016). Nanotechnology contributes much towards the development and introduction of Texture-modified foods (TM), the foods with soft-textures and/or reduced particle size, thickened liquids/drinks like soups, creamy food jellies, mousses, purees, minced and moist foods, and process-softened foods etc. These foods are mainly introduced in the market to overcome the nutritional deficiencies of a senior population with eating dysfunctions (Kiss, 2020). Novel technologies like microfluidics, 3D printing, electrospinning, elecrospraying etc. are used to prepare TM foods which are soft and more nutritious (Gomez- Mascaraque et al., 2015, Godoi et al., 2016; Ghorani and Tucker, 2015; Marquis et al., 2015).

Proteins, carbohydrates, and lipids are the basic components of TM foods. In the case of protein drinks, the globular proteins get unfolded and denatured when heated and increase the viscosity of liquids. These proteins may be self-assembled into nano-sized aggregates and fibrils on further heating and form network chains of gels. Out of carbohydrates, polysaccharides are used to condense aqueous food, stabilize emulsions and foams, and also act as gelling agents. Some of the carbohydrates like gums and starches are used as thickening agents and to encapsulate certain nutrients, antioxidants and enzymes. Lipids are also used abundantly in TM foods because of their amphiphilic nature and tendency to form nano sized micelles and vesicles. Monoglycerides and phospholipids are used as emulsifying agents and to deliver nutrients and other bioactive components at specific targets. Because of their small size and soft texture, these lipids not only modify texture of food but they also elicit stronger aroma feel during breakdown in the mouth (Kiss, 2020).

### 9.2.3 Nanoadditives and Nutraceuticals: Improve Nutritional Value of Food

The bulk of bioactive molecules, such as lipids, proteins, carbohydrates, and vitamins, are sensitive to the stomach and duodenum's acidic environment and enzyme activity. So these bioactive substances are needed to be encapsulated as it enables them to resist adverse acidic conditions and also helps them to assimilate readily along with other food products. Nanoparticles based coating are developed which are used to encapsulate important medicines, vitamins and micronutrients and assist in their delivery to specific organs. The nanocomposite, nano-emulsification, and nanostructuration are the major techniques which are developed to deliver proteins, antioxidants, vitamins and nutrients to provide targeted health benefits. Many flavonoids and vitamins are encapsulated in polymeric based nanoparticles for their safe transport and targeted delivery (Koo et al., 2005; Langer and Peppas, 2003, Singh et al., 2017).

In the area of enriched human food, nanoencapsulation of the vitamins is an emerging field which helps to develop vitamins enriched food. Vitamins are necessary for human health but as they are prone to degradation, their bioavailability during consumption may be limited. But by entrapping these vitamins in the nanocapsules, excellent bioavailability of both water soluble and oil soluble vitamins can be achieved. Nanoencapsulation is a promising approach for targeted delivery and controlled release of vitamins. Various techniques of nanoencapsulation *viz.* nanocarriers like nanoparticles, nanoliposomes, and nanoemulsions etc. not only help in safe delivery but also mask the undesirable flavor and increase their shelf-life by increasing colloidal stability. Oil soluble vitamins like vitamin D, D2, B3, E, E acetate are encapsulated by nanoemulsion, protein nanoparticles, casein micelles, polysaccharides nanoparticles and liposome etc. (Lee et al., 2016; Dasgupta et al., 2016a). Whereas water soluble vitamins like vitamin B9, B2, C, B12 are encapsulated by nanoemulsion, double emulsion, liposomes and small unicellular vesicles etc. (Kiss, 2020; Assadpour et al., 2016). Another group of oil soluble vitamins lycopene, carotene, lutein are encapsulated by "chitosome" (chitosane coated liposomes) (Tan et al., 2016).

## 9.3 ROLE OF NANOTECHNOLOGY IN FOOD PACKAGING

Packaging materials are those products which are used to protect, to contain, to distribute, to transport and to identify any article along its supply chain, from unprocessed

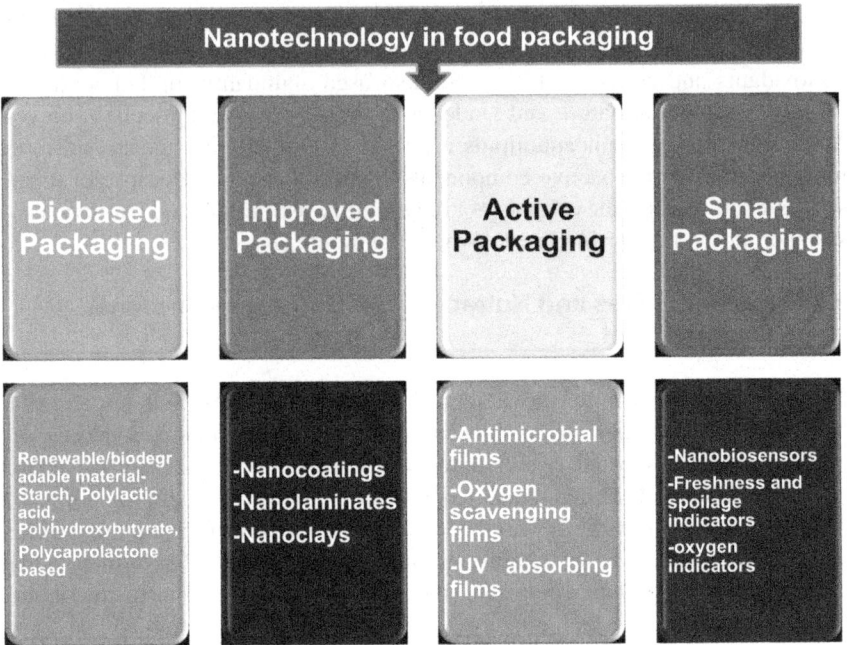

Figure 9.2 Application of nanotechnology in food packaging.

material to end users. The basics conditions for any material to act as packaging material are having decent mechanical, thermal and barrier properties. Various types of food packaging are available which help to increase the shelf life of products and also maintain the quality of product. Nanotechnology is the science of very small materials in the range of nm that has a major role in the food industry including packaging. Various nanomaterials such as silver nanoparticle, nano-zinc oxide, titanium nitride nanoparticle, nano-titanium dioxide and nanoclay are used in food packaging (Pal, 2017; Tager, 2014). It was reported that more than 500 nanopackaging materials are in commercial use for food packaging. In nano packed food material, it is observed that nanomaterial release various anti-microbes, antioxidants, enzymes and different flavors to further enhance the shelf life of food products. Various new techniques, protocols and products related to nanopackaging have come up that have direct application in food and agriculture industry. Nanotechnology facilitated food packaging can be divided into four major categories (Duncan, 2011; Kuswandi, 2016): Bio-based packaging, Active packaging, Smart packaging and Improved packaging (Figure 9.2).

### 9.3.1 Bio-Based Packaging

Non-degradable petroleum-based plastic polymers make up the majority of materials utilized in the packaging industry today. As a result, non-degradable food packaging materials pose a severe environmental threat worldwide (Kirwan and Strawbridge, 2003). So, the adoption of bio-based packaging materials, such as edible and biodegradable films derived from renewable resources (Tharanathan, 2003), could help to alleviate the waste problem by lowering packaging waste and extending shelf life, hence improving food quality. Further, it is possible to maintain product quality and freshness during the time required for commercialization and consumption by using the appropriate packaging materials and technologies (Kuswandi et al., 2011). Currently, however, the usage of bio-based materials for food packaging is quite limited. This is due to the low barrier and mechanical qualities of natural polymers. As a result, natural polymers were commonly combined with synthetic polymers or chemically modified in order to broaden their packaging possibilities. Furthermore, bio-based packaging, like traditional packaging, must perform a variety of critical roles, including food confinement and protection, safety and quality, and providing information to users. Also the application of the nanocomposites assures the enhanced use of edible and biodegradable films in packaging which will help in the preservation of fresh foods by extending their shelf life (Lagaron et al., 2005; Kuswandi, 2016).

Biobased packaging films are thought to be more environmentally friendly than traditional packaging films. Bio-based packaging, like any other type of packaging, creates a barrier between a food product and its surroundings, protecting it from bacteria, environmental relative humidity, and gas conditions. Biodegradable packaging films are distinct from other packaging solutions by their ability to decompose through the action of living organisms. Because the breakdown of products is entirely natural, like carbon dioxide, biomass, and water, this packing method is typically thought to be more environmentally safe. These biodegradable polymers can be categorized based on their origin: (i) Polymers such as polysaccharides, proteins, polypeptides, and polynucleotides that are directly isolated from biomass. (ii) Polymers such as polylactic acid and bio-polyester, which are made by chemical synthesis of bio-based monomers or mixed biomass and petroleum. (iii) Polyhydroxybutyrate, bacterial cellulose, xanthan, curdian, and pullan are examples of polymers generated by microorganisms or genetically modified bacteria (Del Nobile et al., 2009; Kuswandi, 2016).

The major problems associated with biodegradable polymers are their performance, processing and cost. This is because, regardless of origin, all biodegradable polymers have the same "performance and processing". Brittleness, low thermal distortion temperature, high gas and vapor permeability, and poor resistance to corrosion are all issues. Their applications have been severely constrained due to lengthy processing operations. The application of nanomaterials to these biobased polymers could improve their physical properties and also make them cost effective. Three main types of composites are formed when layered clays/silicate are incorporated with the polymer *viz.* tactoid, phase-separated microcomposite; intercalated nanocomposite and exfoliated, polymer clay nanocomposite (Alexandre and Dubois, 2000). These bio-nanocomposites have improved mechanical, thermal and physico-chemical properties as compared to biopolymers. Currently, most suitable bio-nanocomposites for packaging applications are starch and its derivates, i.e. polylactic acid, polybutylene succinate, polyhydroxybutyrate, and aliphatic polyester that can be used for improved packaging and active packaging (Ray and Bousmina, 2005).

### 9.3.2 Improved Packaging

Improved packaging is helpful in barrier performance pertaining to gases such as oxygen, carbon dioxide, and ultraviolet rays, as well as it also adding strength, stiffness, dimensional stability, and heat resistance to packaging. Various nanoparticles/nanocomposites are incorporated with clay nanoparticles in the ratio of 5% (w/w) to increase mechanical and physical properties of the packaging including resistance to temperature and humidity, mechanical strength, and flexibility. This type of packaging is commonly used to pack beer, edible oils and carbonated drinks etc. (Chaudhry et al., 2008). USFDA (United States Food and Drug Administration) has approved the use of nanocomposite in contact with foods (Sozer and Kokini, 2009)

Excess of clay loading in ethylene-vinyl alcohol copolymer based nanocomposite films leads to the reduction of tensile properties and optical transparency due to the formation of clay agglomerates. But with the addition of a lesser amount of clay (3% w/w), the oxygen (by 59%) and water vapour barrier (by 90%), the properties of nanocomposite films were improved in comparison to the material without added montmorillonite clay nanoparticles (Kim and Cha, 2014). Similarly, Arora et al. (2011) reported that 4% (w/w) nano clay leads to an increase in oxygen barrier properties of polystyrene (PS) by 51%. Different nanotechnology methods available for improvement of mechanical and physical properties of the packaging are nano-coatings, nano-laminates and nano-clays.

Nano-coating involves coating of food in the form of thin layer or film, placed or covered on food. These types of coatings provide mass transfer barrier and they can be edible. On the other hand, non-edible nanocoatings are also available which form protective covering of containers. Edible nanocoatings can be hydrocolloids (like polysaccharides with water soluble properties, derivatives of cellulose, starch, pectins, alginates, gum etc.); lipids (like fatty acids, acylglycerols, waxes or essential oils etc.) or proteins (like milk and soy proteins, gelatin, zein proteins and myofibrillar proteins etc.) (Primožič et al., 2021; Wang et al., 2019; Alizadeh-Sani et al., 2018; Wu et al., 2019; Cano et al., 2016). A very thin layer of these edible nanocoatings (5 nm) can be used in the meat processing industry, agriculture industry, cheese and baking industry to maintain freshness of products, to provide flavor, color, enzymes, antioxidants and antibrowing compounds to the products (Nile et al., 2020). By spraying, immersion, or rubbing edible nano-coatings can be easily applied. Ingredients of nano-coatings are environment friendly and they have antimicrobial and antioxidative properties which are helpful to maintain quality of produce (Zambrano-Zaragoza et al., 2018; Yousuf et al., 2018; Wu et al., 2012).

To improve physical characteristics of any food products low cost nano-clay particles are integrated and embedded inside the matrix of packaging material where they provide mechanical and thermal barriers (Majeed et al., 2013; Gabr et al., 2015). In a nanocomposite with included clays, transport of diffusing molecules is obstructed due to impenetrable particles/clays. Nonlinear and prolonged pathway of oxygen and water vapor permeability formed due to the incorporation of clay into a polymer matrix film, which contributes to the prolongation of shelf life of quick spoiled foods. Food packaging materials based on clay nanocomposite offer enhanced shelf life, lightweight, heat resistant and are smash proof (Primožič et al., 2021). Clay montmorillonite/pectines reduce diffusion of water vapor and oxygen in packed food products (Mangiacapra et al., 2006, Yussuf et al., 2018). On the other hand, polycaprolactone/organo nanoclay/chitosan show antimicrobial effect on *E. coli*, *Pseudomonas aeruginosa*, and *Candida albicans* etc. (Cesur et al., 2018).

Nano-laminate films consist of two or more nanolayers which are linked to each other by physical or chemical ways. These films are synthesized by "layer by layer" deposition. This method enables surface lamination with multiple nano-layers as interfacial films based on various nanomaterials. Various adsorbing substances such as natural polyelectrolytes (proteins, polysaccharides), charged lipids (phospholipids, surfactants), and colloidal particles (micelles, vesicles, droplets) are used to form different layers for the preparation of nano-laminates for food packaging (Kuswandi and Moradi, 2019; Dasgupta et al., 2016b). Various antimicrobials, antioxidants, anti-browning agents, enzymes, flavors, odor etc can be incorporated into these films (Galus et al., 2020) to extend the quality and shelf life of packaged food materials (Salgado et al., 2015). Nano-laminated coatings could also be prepared from edible or bio based ingredients as in the case of edible nano-coated films (Primožič et al., 2021). Renewable and cheap biopolymers are used to make biodegradable film for food packaging. Usually polysaccharides such as cellulose, chitosan, starch, pectin, alginate, carrageenan etc. are used for biodegradable coatings. These coatings films have good barrier properties against oxygen and carbon dioxide transport and also show good tensile strength (Cazón et al., 2017; Bhatia, 2016). The tensile strength of amylose starch based films is comparable to low-density polyethylene films (Domene-López et al., 2019). Further plasticizers like polyols (glycerol, sorbitol), sugars (glucose, sucrose) and lipids (mono-, tri-glycerides, phospholipids) are added to improve mechanical strength, flexibility, density, and viscosity of protein and polysaccharides based films (Espitia et al., 2014; Vieira et al., 2011).

### 9.3.3 Active Packaging

In active packaging, nanomaterials are used to interact directly with the food or the environment for the better protection of the packed food products. Several nanoparticles including silver nanoparticles, silver coatings, nano titanium dioxide and carbon nanotubes, nano magnesium oxide and nanocopper oxide act as antimicrobial agents as they absorb oxygen, ethylene gases, UV rays and moisture there by preventing the growth of microbes on packed food. Some of them also release carbon dioxide that create unfavorable environment for microbial growth (Chaudhry et al., 2008; Primožič et al., 2021). Antimicrobial packaging for food products, which absorbs oxygen has been developed and commercialized by Kodak company. Even some of the enzymes present between polyethylene films acting as oxygen scavengers have been developed for antimicrobial packaging (Lopez-Rubio et al., 2006). Further food spoilage represents a serious environmental problem (Mustafa and Andreescu,

2020) and also affects human health. Therefore, the development of new technologies that lead to reduction in food spoilage and wastage is need of the hour. One of the possible strategies for this is development of active packaging that extends the shelf life of food and reduces the wastage and spoilage of food. Traditionally non-biodegradable materials like polyethylene, polypropylene etc. are used for packaging that prevent entry of oxygen and water. But at present new active packaging materials that act as scavengers, absorbers, emitters, and coatings are introduced that are used in conjunction with biodegradable components to improve food quality and safety ( Primožič et al., 2021).

### 9.3.3.1 Antimicrobial Active Packaging/Antimicrobial Films

The purpose of antimicrobial active food packaging is preservation of foods and extending their shelf life by inhibiting the microorganism's growth. This could be achieved either by incorporation of an active agent (nanoparticles) into or by applying a coating layer within the packaging material (Mustafa and Andreescu, 2020). Antimicrobial agents are based on different physiological requirements/ survival strategies of microorganisms. Different microbes responsible for food spoilage are *Salmonella spp.*, *Staphylococcus aureus*, *Listeria monocytogenes*, *Bacillus cereus*, *Escherichia coli*, *Pseudomonas*, *Klebsiella*, *Lactobacillus spp.*; *Rhizopus*, *Aspergillus* (molds); and *Torulopsis*, *Candida* (yeasts) (Vilela et al., 2018). Antimicrobial agents either inhibit the essential metabolic pathways of microorganisms or destroy their cell wall/membrane to hamper their growth.

Nanomaterials are introduced to the food packaging system due to their antimicrobial, UV protection activity, and oxygen absorbing activity etc. Ag, $TiO_2$, ZnO and MgO nanoparticles are common antibacterial nanoparticles which are extensively utilized in antibacterial films of packaging (Becerril et al., 2020). $TiO_2$ nanoparticles are recognized as non-toxic to human body and approved as a food additive and for use in food packaging (Chaudhary et al., 2020). Metal-based nanoparticles in combination with other essential oils, polyphenols etc. by nanoencapsulation techniques are used as effective antimicrobial agents for active food packaging (Valdes et al., 2015; Rehman et al., 2020). Plasticized polylactide (PLA) silver-copper (Ag-Cu) nanoparticles and cinnamon essential oil (CEO) film was found to be effective for packaging of chicken meat. The composite films of nanoparticles along with essential oil show antibacterial activity against *Salmonella Typhimurium*, *Campylobacter jejuni*, and *L. monocytogenes* for 21 days when chicken samples were stored in the refrigerator (Ahmed et al., 2018). Buckwheat starch (BS) films containing ZnO nanoparticles showed great antimicrobial activity against *L. monocytogenes* for freshly-cut mushroom packaging (Kim and Song, 2018). $TiO_2$/ZnO nanoparticles-coated low-density polyethylene (LDPE) films in the presence of UV light reduce *E. coli* growth in packed meat (Marcous et al., 2017). Antibacterial polyvinyl alcohol-based nanocomposite film prepared with Montmorillonite clay and ginger extract mediated Ag nanoparticles exhibit antimicrobial activity against both *S. Typhimurium* (Gram-negative bacteria) and *S. aureus* (Gram-positive bacteria) (Mathew et al., 2019). Various polyethylene films containing Ag, clay, and TiO2 nanoparticles were designed to use as a potential active packaging film with the aim to enhance fresh chicken shelf life stored at 4⁰C. The film containing 5% Ag and 5% TiO$_2$ nanoparticles was observed to show antimicrobial effect on gram-positive (*S. aureus*) and gram-negative bacteria (*E. coli*) (Lotfi et al., 2019).

Nanocomposite poly (ethylene oxide) films with Ag nanoparticles and extract of *Acca sellowiana* demonstrated antimicrobial activity against *E. coli* and *S. aureus* (Sganzerla et al., 2020). Biodegradable chitosan-whey protein-based film containing TiO$_2$ and *Zataria multiflora* essential oil also exhibited strong antimicrobial properties against *S. aureus, E. coli, and L. monocytogenes*. Even the gelatin-based nanocomposite with incorporated chitosan nanofiber and ZnO nanoparticles were found to be effective for active packaging of chicken fillet and cheese. (Amjadi et al., 2019). Hydroxypropyl methylcellulose/beeswax biocomposite film with Ag nanoparticles was found to be effective even against pathogenic bacteria in packed food. Biocomposite film inhibited the growth of gram-positive bacteria (*B. cereus, S. aureus, Streptococcus pneumoniae,* and *L. monocytogenes*) and gram-negative bacteria (*E. coli, S. typhimorum, P. aeruginosa,* and *Klebsiella pneumoniae*) in a dose-dependent manner (Bahrami et al., 2019). Polyethylene coated with chitosan-ZnO nanocomposite films also shows very high antimicrobial activity by completely inhibiting the growth of food pathogens such as *Salmonella enterica, E. coli,* and *S. aureus* (Al-Naamani et al., 2016). Antimicrobial activity of ZnO nanoparticles was further confirmed by Rezaei et al. (2020). It was observed that pure gluten film possessed no antimicrobial activity against *E. coli* and *Aspergillus niger*, but films with incorporated ZnO nanoparticles showed a significant antimicrobial activity against bacteria and fungi. Nanocomposite films with encapsulated bioactive compounds like curcumin were found to show antioxidant (radical scavenging) properties and antibacterial activity against *E. coli* (Valencia et al., 2019). Active food packaging not only show antimicrobial properties but it also displays enhanced thermal, physicochemical, mechanical, and optical properties of food packaging (Jariyasakoolroj et al., 2020; Atares and Chiralt, 2016).

### 9.3.3.2 Oxygen Scavenging Films

Oxygen is responsible for the deterioration of many foods either directly or indirectly. Direct oxidation of food results in browning of fruits and rancidity of vegetable oils. On the other hand, indirect oxidation of food includes food spoilage by aerobic microorganisms in the presence of oxygen. So oxygen scavengers in the form of films containing titania nanoparticles introduced to different polymers are acquainted into food packaging industry to maintain a low level of oxygen which further enhances shelf-life of the food (Kuswandi, 2016). TiO$_2$ helps to eliminate oxygen by photocatalytic mechanisms in the presence of ultraviolet radiation (Mills et al., 2006).

### 9.3.3.3 UV Absorbing Films

Nanocrystalline titania (TiO$_2$) is commonly used as UV absorbing film. TiO$_2$ coated films exposed to sunlight inactivate fecal coli forms in water (Gelover et al., 2006).

Further metal fixing improves visible light absorbance capacity of $TiO_2$ and increases its photo catalytic efficiency under UV irradiation (Anpo et al., 2001). It has been reported that doping of $TiO_2$ with silver greatly improved photo catalytic efficiency which cause bacterial inactivation (Reddy et al., 2007). So, combination of $TiO_2$ with silver nanoparticles in a nanocomposite has good antibacterial properties (Cheng et al., 2006).

### 9.3.4 Smart/Intelligent Packaging

Smart packaging is designed for sensing biochemical or microbial changes in the food. It can detect the development of specific pathogen or specific gases arising from the food spoilage. Some smart packaging has been developed to use as tracing device for food safety. Currently, various food companies and super markets including Nestle, MonoPrix Super market and British Airways are using chemical sensors, which can easily detect color change due to food spoilage (Pal, 2017). This smart packaging is possible with the use of latest techniques of nanotechnology which are used to monitor biochemical changes and microbial development inside the food and or the environment surrounding the product (Kuswandi, 2017; Madhusudan et al., 2018). Due to specific optical properties, photocatalytic activities and high surface reactivity of nanomaterials as compared to traditional colorimetric indicator, nanomaterial are used as nanosensors to detect damage in food products.

#### 9.3.4.1 Bionanosensors

Nanosensors as optical indicators help to ensure food quality and safety by providing information to customer about changes in internal or external parameter of the food and/or in its surrounding environment. For this different nanomaterial like $SiO_2$ nanoparticles, colloid gold nanoparticles, single walled and multi walled carbon nanotubes, Poly (amidoamine) dendrimers, $Fe_3O_4$ magnetic gold nanoparticles and graphene nanoplatelets are used to improve and upgrade the functionality of packaging (Bumbudsanpharoke, and Ko, 2019; Primožič et al., 2021). Nanosensors have great potential for fast detection, identification, and quantification of pathogenic microorganisms, decaying substances, and allergy-causing proteins. E custom-made nanosensors used in smart packaging for food analysis including detection of toxins, chemicals, and food pathogens and detection of flavors or colors etc. (Li and Sheng, 2014). Food packaging equipped with nanosensors can also detect changes in level of humidity, temperature, gas formation etc. The changes in level of these parameters are because of food spoilage which in turn changes the color of indicator and thus alerting the customer to the unsuitability of the product. So, these types of food packaging with nanosensors, can be successfully used for real-time monitoring of food freshness status, and reduce the requirement for determining the expiry date of the food (Sharma et al., 2017; Pramanik et al., 2020).

A biosensor is a device with a biological sensing element connected to a transducer. Biological components are analyzed with the help of biosensors. Bionanosensors are developed by combining the biosensors with nanomaterials which have great usage in food industry. In addition to checking the quality and level of pathogens in food, they can also detect the presence of pesticides and chemicals in food products (Wu et al., 2013; Kara et al., 2013; Majidi et al., 2013). $SiO_2$ nanomaterial based bionanosensors are used to detect the presence of mycotoxins in packed food products, whereas colloidal gold nanoparticles based bionanosensors sensed the presence of botulinum neurotoxin or brevetoxins (Zhou et al., 2009; Actis et al., 2010). In addition to this several nanomaterials based bionanosensors containing gold nanoparticles, graphene nanoplatelet, $Fe_3O_4$ magnetic gold nanoparticles, carbon nanotubes etc. detect the presence of various pathogenic microbes *viz. E. coli, L. monocytogenes, S. enterica, S. aureus*, and *S. typhimurium* (Wang et al., 2015; Zhang et al., 2015; Pandey et al., 2017).

### *9.3.4.2 Oxygen Indicators*

In the presence of oxygen aerobic microorganisms contaminate the food during storage. So it is needed to develop non-toxic and irreversible oxygen sensors that assure oxygen absence in oxygen free food packaging systems, such as packaging under vacuum or nitrogen. $TiO_2$ nanoparticles based UV-activated colorimetric oxygen indicator using UVA light has been developed. $TiO_2$ nanoparticles photosensitize the reduction of methylene blue by triethanolamine in a polymer encapsulation medium. Upon UV irradiation, the sensor remains colorless, until it is exposed to oxygen, where its original blue color is restored. The rate of development of blue color is proportional to the level of oxygen exposure (Lee et al. 2005). Similarly another nanoparticles based oxygen indicator is developed to indicate the level of $O_2$ in packed food. In this case nano crystalline $SnO_2$ has been used as a photo sensitizer in a colorimetric $O_2$ indicator in combination with electron donor -glycerol, a redox dye -methylene blue, and an encapsulating polymer -hydroxyethyl cellulose (Mills and Hazafy, 2009). In the presence of UVB light the oxygen indicator get activated (photo bleached) and $SnO_2$ nanoparticles reduced methylene blue dye. In the presence of $O_2$ in packed food the color of the nanoparticles films changed to blue which was otherwise colourless.

## 9.4 NANOTECHNOLOGY AND FOOD SAFETY ISSUES

No doubt there are lots of advantages of nanotechnology to the food industry as it improves processing and packaging of food thereby reducing the wastage and spoilage of food. But safety issues regarding human health and environment associated with nanoparticles cannot be ignored. Many researchers proposed that nanoparticles may be migrated from packaging material and impact on health of consumers (Jain et al., 2018; Singh et al., 2017). Nanoparticles both inorganic (like silver, iron oxide, titanium dioxide, silicon dioxide, and zinc oxide) and organic (like lipid, protein, and carbohydrate) in foods may have toxic effects. Humans are exposed to NPs through various means including oral intake of NPs-containing food or beverage. Other means to exposure to NP are skin and inhalation, so safety aspects of NPs to food industry duly need a consideration (Go et al., 2017). The nanoparticles may behave differently within the human body owing to their smaller size as compared to the larger particles usually utilized as food ingredients. The possible toxicity of nanoparticles is determined by their rate of absorption, distribution, metabolism and excretion. If any ingested nanoparticle not digested in upper gastrointestinal tract, may reach the lower gastrointestinal tract where it may alter the gut microbiota (Baliyan et al., 2020; McClements and Xiao, 2017;). Nanoparticle's behavior, fate, transport and toxicity influenced by their physical factors *viz.* size, shape, surface charge, surface area, coating, crystal structure and solubility; and external factors like temperature, pH, ionic strength, salinity, and presence of organic matter etc. Nanoparticles may produce toxicity in cells through a variety of the ways. One of the main mechanisms by which they induce toxicity in cells is production of reactive oxygen species (ROS) just like their macro-counterparts. Excess of the ROS (superoxide radicals, singlet oxygen, hydrogen peroxide and hydroxyl radicals) produced under the effect of nanoparticles interact with important structural biomolecules like lipids, proteins, or nucleic acids by disturbing the structure of cell membranes, nucleus and various other cell organelles. These alterations at structural, biochemical and genomic levels may be fatal or cancerous to cells (Sharma et al., 2014).

Although the material which is considered as GRAS (generally regarded as safe) substance, additional studies must be carried to examine the risk of its nano corresponds as the physico-chemical properties of any components in its nanostates are completely different from their macrostate. In addition to this nanosized particles have higher tendency to accumulate in body tissues (Savolainen et al., 2010). In one of the study made by Athinarayanan et al. (2014), it was observed that silica nanoparticles which are used for food processing and packaging can be cytotoxic to human lung cell lines. Nanoparticles like Ag and Cu can migrate from packaging material to food. The rate of migration of nanoparticles is determined by the percentage of nanofiller in the nanocomposites (Cushen et al., 2014). So regulatory authorities must propose some standards for use of nanoparticles in food processing and the food packaging industry to ensure quality of products delivered to consumers, health and wellbeing of consumers and for safety of environment.

## 9.5 CONCLUSION

Processing and packaging of food material ensure the delivery of products to consumer in better conditions for their final use. Additionally food processing using nanomaterials mask the unwanted flavors/odors, improve the texture and consistency of food and enhanced the nutraceutical value of food products by stabilizing the important vitamins and other micronutrients. The use of protective coatings and other suitable packaging using nanomaterial based technology not only increase the shelf life of food products but also enhanced their nutritional values. Intelligent food packaging using nanobiosensors is boon for framers as well as for consumers as it provides necessary information to consumer about nutritional status, level of microbes and level of gases (oxygen, carbon dioxide and ethylene) there by making the safe and hygienic use of food products. But it should be imperative to test nanoparticles processed foods before they are released in market. So, insight of mechanisms involved in interaction of nanoparticles with food and biological systems will help to develop better technology where nanoparticles will be used in appropriate way for safer food packaging and processing.

## REFERENCES

Abbas, K. A., Saleh, A. M., Mohamed, A., & MohdAzhan, N. (2009). The recent advances in the nanotechnology and its applications in food processing: a review. *J Food Agric Environ*, 7(3–4), 14–17.

Actis, P., Jejelowo, O., & Pourmand, N. (2010). Ultrasensitive mycotoxin detection by STING sensors. *Biosensors and Bioelectronics*, 26(2), 333–337.

Aguilera, J. M., & Park, D. J. (2016). Texture-modified foods for the elderly: Status, technology and opportunities. *Trends in Food Science & Technology*, 57, 156–164.

Ahmed, J., Arfat, Y. A., Bher, A., Mulla, M., Jacob, H., & Auras, R. (2018). Active chicken meat packaging based on polylactide films and bimetallic Ag–Cu nanoparticles and essential oil. *Journal of Food Science*, 83(5), 1299–1310.

Alexandre, M., & Dubois, P. (2000). Polymer-layered silicate nanocomposites: preparation, properties and uses of a new class of materials. *Materials Science and Engineering: R: Reports*, 28(1–2), 1–63.

Alizadeh-Sani, M., Khezerlou, A., & Ehsani, A. (2018). Fabrication and characterization of the bionanocomposite film based on whey protein biopolymer loaded with TiO2 nanoparticles, cellulose nanofibers and rosemary essential oil. *Industrial Crops and Products*, 124, 300–315.

Al-Naamani, L., Dobretsov, S., & Dutta, J. (2016). Chitosan-zinc oxide nanoparticle composite coating for active food packaging applications. *Innovative Food Science & Emerging Technologies*, 38, 231–237.

Amjadi, S., Emaminia, S., Nazari, M., Davudian, S. H., Roufegarinejad, L., & Hamishehkar, H. (2019). Application of reinforced ZnO nanoparticle-incorporated gelatin bionanocomposite film with chitosan nanofiber for packaging of chicken fillet and cheese as food models. *Food and Bioprocess Technology*, 12(7), 1205–1219.

Anpo, M., Kishiguchi, S., Ichihashi, Y., Takeuchi, M., Yamashita, H., Ikeue, K., ... & Che, M. (2001). The design and development of second-generation titanium oxide photocatalysts able to operate under visible light irradiation by applying a metal ion-implantation method. *Research on Chemical Intermediates*, 27, 459–467.

Arora, A., Choudhary, V., & Sharma, D. K. (2011). Effect of clay content and clay/surfactant on the mechanical, thermal and barrier properties of polystyrene/organoclay nanocomposites. *Journal of Polymer Research*, 18(4), 843–857.

Assadpour, E., Maghsoudlou, Y., Jafari, S. M., Ghorbani, M., & Aalami, M. (2016). Optimization of folic acid nano-emulsification and encapsulation by maltodextrin-whey protein double emulsions. *International Journal of Biological Macromolecules*, 86, 197–207.

Atarés, L., & Chiralt, A. (2016). Essential oils as additives in biodegradable films and coatings for active food packaging. *Trends in Food Science & Technology*, 48, 51–62.

Athinarayanan, J., Periasamy, V. S., Alsaif, M. A., Al-Warthan, A. A., & Alshatwi, A. A. (2014). Presence of nanosilica (E551) in commercial food products: TNF-mediated oxidative stress and altered cell cycle progression in human lung fibroblast cells. *Cell Biology and Toxicology, 30*(2), 89–100.

Bahrami, A., Mokarram, R. R., Khiabani, M. S., Ghanbarzadeh, B., & Salehi, R. (2019). Physico-mechanical and antimicrobial properties of tragacanth/hydroxypropyl methylcellulose/beeswax edible films reinforced with silver nanoparticles. *International Journal of Biological Macromolecules, 129*, 1103–1112.

Baliyan, N., Rani, R., Kaur, P., Yadava, Y. K., & Kumar, L. (2020). Nanoencapsulation Development for Interactive Foods. *Chem Sci Rev Lett, 9*, 1039–1057.

Becerril, R., Nerín, C., & Silva, F. (2020). Encapsulation systems for antimicrobial food packaging components: An update. *Molecules, 25*(5), 1134.

Bhatia, S. (2016). Natural polymers vs synthetic polymer In: *Natural Polymer Drug Delivery Systems: Nanoparticles, Plants, and Algae* (pp. 95–118). Springer. Cham, Switzerland. ISBN 978-3-319-41129-3.

Bumbudsanpharoke, N., & Ko, S. (2019). Nanomaterial-based optical indicators: Promise, opportunities, and challenges in the development of colorimetric systems for intelligent packaging. *Nano Research, 12*(3), 489–500.

Cano, A., Cháfer, M., Chiralt, A., & González-Martínez, C. (2016). Development and characterization of active films based on starch-PVA, containing silver nanoparticles. *Food Packaging and Shelf Life, 10*, 16–24.

Cazón, P., Velazquez, G., Ramírez, J. A., & Vázquez, M. (2017). Polysaccharide-based films and coatings for food packaging: A review. *Food Hydrocolloids, 68*, 136–148.

Cesur, S., Köroğlu, C., & Yalçın, H. T. (2018). Antimicrobial and biodegradable food packaging applications of polycaprolactone/organo nanoclay/chitosan polymeric composite films. *Journal of Vinyl and Additive Technology, 24*(4), 376–387.

Chaudhary, P., Fatima, F., & Kumar, A. (2020). Relevance of nanomaterials in food packaging and its advanced future prospects. *Journal of Inorganic and Organometallic Polymers and Materials, 30*(12), 5180–5192.

Chaudhry, Q., Scotter, M., Blackburn, J., Ross, B., Boxall, A., Castle, L., ... & Watkins, R. (2008). Applications and implications of nanotechnologies for the food sector. *Food Additives and Contaminants, 25*(3), 241–258.

Cheng, Q., Li, C., Pavlinek, V., Saha, P., & Wang, H. (2006). Surface-modified antibacterial TiO2/Ag+ nanoparticles: Preparation and properties. *Applied Surface Science, 252*(12), 4154–4160.

Cushen, M., Kerry, J., Morris, M., Cruz-Romero, M., & Cummins, E. (2014). Evaluation and simulation of silver and copper nanoparticle migration from polyethylene nanocomposites to food and an associated exposure assessment. *Journal of Agricultural and Food Chemistry, 62*(6), 1403–1411.

Dasgupta, N., Ranjan, S., Mundekkad, D., Ramalingam, C., Shanker, R., & Kumar, A. (2015). Nanotechnology in agro-food: from field to plate. *Food Research International, 69*, 381–400.

Dasgupta, N., Ranjan, S., Mundra, S., Ramalingam, C., & Kumar, A. (2016a). Fabrication of food grade vitamin E nanoemulsion by low energy approach, characterization and its application. *International Journal of Food Properties, 19*(3), 700–708.

Dasgupta, N., Ranjan, S., Patra, D., Srivastava, P., Kumar, A., & Ramalingam, C. (2016b). Bovine serum albumin interacts with silver nanoparticles with a "side-on" or "end on" conformation. *Chemico-biological Interactions, 253*, 100–111.

de Vos, P., Faas, M. M., Spasojevic, M., & Sikkema, J. (2010). Encapsulation for preservation of functionality and targeted delivery of bioactive food components. *International Dairy Journal, 20*(4), 292–302.

Dekkers, S., Krystek, P., Peters, R. J., Lankveld, D. P., Bokkers, B. G., van Hoeven-Arentzen, P. H., ... & Oomen, A. G. (2011). Presence and risks of nanosilica in food products. *Nanotoxicology, 5*(3), 393–405.

Del Nobile, M. A., Conte, A., Buonocore, G. G., Incoronato, A. L., Massaro, A., & Panza, O. (2009). Active packaging by extrusion processing of recyclable and biodegradable polymers. *Journal of Food Engineering, 93*(1), 1–6.

Desai, K. G. H., & Jin Park, H. (2005). Recent developments in microencapsulation of food ingredients. *Drying Technology, 23*(7), 1361–1394.

Domene-López, D., García-Quesada, J. C., Martin-Gullon, I., & Montalbán, M. G. (2019). Influence of starch composition and molecular weight on physicochemical properties of biodegradable films. *Polymers, 11*(7), 1084.

Duncan, T. V. (2011). Applications of nanotechnology in food packaging and food safety: barrier materials, antimicrobials and sensors. *Journal of Colloid and Interface Science, 363*(1), 1–24.

Espitia, P. J. P., Du, W. X., de Jesús Avena-Bustillos, R., Soares, N. D. F. F., & McHugh, T. H. (2014). Edible films from pectin: Physical-mechanical and antimicrobial properties-A review. *Food Hydrocolloids, 35,* 287–296.

Ezhilarasi, P. N., Karthik, P., Chhanwal, N., & Anandharamakrishnan, C. (2013). Nanoencapsulation techniques for food bioactive components: a review. *Food and Bioprocess Technology, 6*(3), 628–647.

Fang, Z., & Bhandari, B. (2010). Encapsulation of polyphenols–a review. *Trends in Food Science & Technology, 21*(10), 510–523.

Fathi, M., Martín, Á., & McClements, D. J. (2014). Nanoencapsulation of food ingredients using carbohydrate based delivery systems. *Trends in Food Science & Technology, 39*(1), 18–39.

Gabr, M. H., Okumura, W., Ueda, H., Kuriyama, W., Uzawa, K., & Kimpara, I. (2015). Mechanical and thermal properties of carbon fiber/polypropylene composite filled with nano-clay. *Composites Part B: Engineering, 69,* 94–100.

Galus, S., Arik Kibar, E. A., Gniewosz, M., & Kraśniewska, K. (2020). Novel materials in the preparation of edible films and coatings—A review. *Coatings, 10*(7), 674.

Gaonkar, A. G., Vasisht, N., Khare, A. R., & Sobel, R. (Eds.). (2014). *Microencapsulation in the Food Industry: A Practical Implementation Guide.* Elsevier Science, Amsterdam, Netherlands.

Gelover, S., Gómez, L. A., Reyes, K., & Leal, M. T. (2006). A practical demonstration of water disinfection using TiO2 films and sunlight. *Water Research, 40*(17), 3274–3280.

Ghorani, B., & Tucker, N. (2015). Fundamentals of electrospinning as a novel delivery vehicle for bioactive compounds in food nanotechnology. *Food Hydrocolloids, 51,* 227–240.

Go, M. R., Bae, S. H., Kim, H. J., Yu, J., & Choi, S. J. (2017). Interactions between food additive silica nanoparticles and food matrices. *Frontiers in Microbiology, 8,* 1013.

Godoi, F. C., Prakash, S., & Bhandari, B. R. (2016). 3d printing technologies applied for food design: Status and prospects. *Journal of Food Engineering, 179,* 44–54.

Gómez-Mascaraque, L. G., Lagarón, J. M., & López-Rubio, A. (2015). Electrosprayed gelatin submicroparticles as edible carriers for the encapsulation of polyphenols of interest in functional foods. *Food Hydrocolloids, 49,* 42–52.

Gupta, A., Eral, H. B., Hatton, T. A., & Doyle, P. S. (2016). Nanoemulsions: formation, properties and applications. *Soft Matter, 12*(11), 2826–2841.

Jain, A., Ranjan, S., Dasgupta, N., & Ramalingam, C. (2018). Nanomaterials in food and agriculture: an overview on their safety concerns and regulatory issues. *Critical Reviews in Food Science and Nutrition, 58*(2), 297–317.

Jariyasakoolroj, P., Leelaphiwat, P., & Harnkarnsujarit, N. (2020). Advances in research and development of bioplastic for food packaging. *Journal of the Science of Food and Agriculture, 100*(14), 5032–5045.

Kara, M., Uzun, L., Kolayli, S., & Denizli, A. (2013). Combining molecular imprinted nanoparticles with surface plasmon resonance nanosensor for chloramphenicol detection in honey. *Journal of Applied Polymer Science, 129*(4), 2273–2279.

Kim, S. W., & Cha, S. H. (2014). Thermal, mechanical, and gas barrier properties of ethylene–vinyl alcohol copolymer-based nanocomposites for food packaging films: Effects of nanoclay loading. *Journal of Applied Polymer Science, 131*(11).

Kim, S., & Song, K. B. (2018). Antimicrobial activity of buckwheat starch films containing zinc oxide nanoparticles against Listeria monocytogenes on mushrooms. *International Journal of Food Science & Technology, 53*(6), 1549–1557.

Kirwan, M. J., & Strawbridge, J. W. (2003). Plastics in food packaging. *Food Packaging Technology, 1,* 174–240.

Kiss, É. (2020). Nanotechnology in food systems: A review. *Acta Alimentaria, 49*(4), 460–474.

Koo, O. M., Rubinstein, I., & Onyuksel, H. (2005). Role of nanotechnology in targeted drug delivery and imaging: a concise review. *Nanomedicine: Nanotechnology, Biology and Medicine, 1*(3), 193–212.

Kuswandi, B. (2016). Nanotechnology in food packaging. In *Nanoscience in Food and Agriculture 1* (pp. 151–183). Springer, Cham.

Kuswandi, B. (2017). Environmental friendly food nano-packaging. *Environmental Chemistry Letters, 15*(2), 205–221.

Kuswandi, B., & Moradi, M. (2019). Improvement of food packaging based on functional nanomaterial. In *Nanotechnology: Applications in Energy, Drug and Food* (pp. 309–344). Springer, Cham.

Kuswandi, B., Wicaksono, Y., Abdullah, A., Heng, L. Y., & Ahmad, M. (2011). Smart packaging: sensors for monitoring of food quality and safety. *Sensing and Instrumentation for Food Quality and Safety, 5*(3), 137–146.

Lagaron, J. M., Cabedo, L., Cava, D., Feijoo, J. L., Gavara, R., & Gimenez, E. (2005). Improving packaged food quality and safety. Part 2: Nanocomposites. *Food Additives and Contaminants, 22*(10), 994–998.

Langer, R., & Peppas, N. A. (2003). Advances in biomaterials, drug delivery, and bionanotechnology. *AIChE Journal, 49*(12), 2990–3006.

Lee, H., Yildiz, G., Dos Santos, L. C., Jiang, S., Andrade, J. E., Engeseth, N. J., & Feng, H. (2016). Soy protein nano-aggregates with improved functional properties prepared by sequential pH treatment and ultrasonication. *Food Hydrocolloids, 55,* 200–209.

Lee, S. K., Sheridan, M., & Mills, A. (2005). Novel UV-activated colorimetric oxygen indicator. *Chemistry of Materials, 17*(10), 2744–2751.

Li, Z., & Sheng, C. (2014). Nanosensors for food safety. *Journal of Nanoscience and Nanotechnology, 14*(1), 905–912.

Lopez-Rubio, A., Gavara, R., & Lagaron, J. M. (2006). Bioactive packaging: turning foods into healthier foods through biomaterials. *Trends in Food Science & Technology, 17*(10), 567–575.

Lotfi, S., Ahari, H., & Sahraeyan, R. (2019). The effect of silver nanocomposite packaging based on melt mixing and sol–gel methods on shelf life extension of fresh chicken stored at 4 C. *Journal of Food Safety, 39*(3), e12625.

Ludwig, C. J., Gaonkar, A. G., & Frey, C. R. (2006). *U.S. Patent No. 7,122,215.* U.S. Patent and Trademark Office, Washington, DC.

Madhusudan, P., Chellukuri, N., & Shivakumar, N. (2018). Smart packaging of food for the 21st century–A review with futuristic trends, their feasibility and economics. *Materials Today: Proceedings*, 5(10), 21018–21022.

Majeed, K., Jawaid, M., Hassan, A. A. B. A. A., Bakar, A. A., Khalil, H. A., Salema, A. A., & Inuwa, I. (2013). Potential materials for food packaging from nanoclay/natural fibres filled hybrid composites. *Materials & Design*, 46, 391–410.

Majidi, M. R., Fadakar Bajeh Baj, R., & Naseri, A. (2013). Carbon nanotube–ionic liquid (CNT–IL) nancamposite modified sol-gel derived carbon-ceramic electrode for simultaneous determination of sunset yellow and tartrazine in food samples. *Food Analytical Methods*, 6(5), 1388–1397.

Mangiacapra, P., Gorrasi, G., Sorrentino, A., & Vittoria, V. (2006). Biodegradable nanocomposites obtained by ball milling of pectin and montmorillonites. *Carbohydrate Polymers*, 64(4), 516–523.

Marcous, A., Rasouli, S., & Ardestani, F. (2017). Low-density polyethylene films loaded by titanium dioxide and zinc oxide nanoparticles as a new active packaging system against Escherichia coli O157: H7 in fresh calf minced meat. *Packaging Technology and Science*, 30(11), 693–701.

Marquis, M., Davy, J., Cathala, B., Fang, A., & Renard, D. (2015). Microfluidics assisted generation of innovative polysaccharide hydrogel microparticles. *Carbohydrate Polymers*, 116, 189–199.

Mathew, S., Snigdha, S., Mathew, J., & Radhakrishnan, E. K. (2019). Biodegradable and active nanocomposite pouches reinforced with silver nanoparticles for improved packaging of chicken sausages. *Food Packaging and Shelf Life*, 19, 155–166.

McClements, D. J., & Xiao, H. (2017). Is nano safe in foods? Establishing the factors impacting the gastrointestinal fate and toxicity of organic and inorganic food-grade nanoparticles. *NPJ Science of Food*, 1(1), 1–13.

Mills, A., & Hazafy, D. (2009). Nanocrystalline SnO2-based, UVB-activated, colourimetric oxygen indicator. *Sensors and Actuators B: Chemical*, 136(2), 344–349.

Mills, A., Doyle, G., Peiro, A. M., & Durrant, J. (2006). Demonstration of a novel, flexible, photocatalytic oxygen-scavenging polymer film. *Journal of Photochemistry and Photobiology A: Chemistry*, 177(2–3), 328–331.

Mozafari, M. R. (2005). Liposomes: an overview of manufacturing techniques. *Cellular and Molecular Biology Letters*, 10(4), 711.

Mustafa, F., & Andreescu, S. (2020). Nanotechnology-based approaches for food sensing and packaging applications. *RSC Advances*, 10(33), 19309–19336.

Nakagawa, K. (2014). Nano-and Microencapsulation of Flavor in Food Systems. *Nano-and Microencapsulation for Foods*, 249–271.

Nile, S. H., Baskar, V., Selvaraj, D., Nile, A., Xiao, J., & Kai, G. (2020). Nanotechnologies in food science: applications, recent trends, and future perspectives. *Nano-micro Letters*, 12(1), 1–34.

Norris, J. T., Zibara, R., Payne, M., Richardson, J., Crawford-Phillips, J., & Dawson Sr, D. J. (2011). *U.S. Patent No. 8,042,356*. U.S. Patent and Trademark Office, Washington, DC.

Ozturk, B., Argin, S., Ozilgen, M., & McClements, D. J. (2015). Formation and stabilization of nanoemulsion-based vitamin E delivery systems using natural biopolymers: Whey protein isolate and gum arabic. *Food Chemistry*, 188, 256–263.

Pal, M. (2017). Nanotechnology: a new approach in food packaging. *J. Food Microbiol. Saf. Hyg*, 2(02), 8–9.

Pandey, A., Gurbuz, Y., Ozguz, V., Niazi, J. H., & Qureshi, A. (2017). Graphene-interfaced electrical biosensor for label-free and sensitive detection of foodborne pathogenic E. coli O157: H7. *Biosensors and Bioelectronics*, *91*, 225–231.

Pathakoti, K., Manubolu, M., & Hwang, H. M. (2017). Nanostructures: Current uses and future applications in food science. *Journal of Food and Drug Analysis*, *25*(2), 245–253.

Pradhan, N., Singh, S., Ojha, N., Shrivastava, A., Barla, A., Rai, V., & Bose, S. (2015). Facets of nanotechnology as seen in food processing, packaging, and preservation industry. *BioMed Research* International, *2015*, 1–18.

Pramanik, P. K. D., Solanki, A., Debnath, A., Nayyar, A., El-Sappagh, S., & Kwak, K. S. (2020). Advancing modern healthcare with nanotechnology, nanobiosensors, and internet of nano things: Taxonomies, applications, architecture, and challenges. *IEEE Access*, *8*, 65230–65266.

Primožič, M., Knez, Ž., & Leitgeb, M. (2021). (Bio) Nanotechnology in food science—food packaging. *Nanomaterials*, *11*(2), 292.

Ray, S. S., & Bousmina, M. (2005). Biodegradable polymers and their layered silicate nanocomposites: in greening the 21st century materials world. *Progress in Materials Science*, *50*(8), 962–1079.

Reddy, M. P., Venugopal, A., & Subrahmanyam, M. (2007). Hydroxyapatite-supported Ag–TiO2 as Escherichia coli disinfection photocatalyst. *Water Research*, *41*(2), 379–386.

Rehman, A., Jafari, S. M., Aadil, R. M., Assadpour, E., Randhawa, M. A., & Mahmood, S. (2020). Development of active food packaging via incorporation of biopolymeric nanocarriers containing essential oils. *Trends in Food Science & Technology*, *101*, 106–121.

Rezaei, M., Pirsa, S., & Chavoshizadeh, S. (2020). Photocatalytic/antimicrobial active film based on wheat gluten/ZnO nanoparticles. *Journal of Inorganic and Organometallic Polymers and Materials*, *30*(7), 2654–2665.

Salgado, P. R., Ortiz, C. M., Musso, Y. S., Di Giorgio, L., & Mauri, A. N. (2015). Edible films and coatings containing bioactives. *Current Opinion in Food Science*, *5*, 86–92.

Santos, D. T., Albarelli, J. Q., Beppu, M. M., & Meireles, M. A. A. (2013). Stabilization of anthocyanin extract from jabuticaba skins by encapsulation using supercritical CO2 as solvent. *Food Research International*, *50*(2), 617–624.

Savolainen, K., Pylkkänen, L., Norppa, H., Falck, G., Lindberg, H., Tuomi, T., ... & Seipenbusch, M. (2010). Nanotechnologies, engineered nanomaterials and occupational health and safety–A review. *Safety Science*, *48*(8), 957–963.

Sawai, J. (2003). Quantitative evaluation of antibacterial activities of metallic oxide powders (ZnO, MgO and CaO) by conductimetric assay. *Journal of Microbiological Methods*, *54*(2), 177–182.

Sekhon, B. S. (2010). Food nanotechnology–an overview. *Nanotechnol Sci Appl*, *3*, 1–15.

Sganzerla, W. G., Longo, M., de Oliveira, J. L., da Rosa, C. G., de Lima Veeck, A. P., de Aquino, R. S., ... & Nunes, M. R. (2020). Nanocomposite poly (ethylene oxide) films functionalized with silver nanoparticles synthesized with Acca sellowiana extracts. *Colloids and Surfaces A: Physicochemical and Engineering Aspects*, *602*, 125125.

Sharma, C., Dhiman, R., Rokana, N., & Panwar, H. (2017). Nanotechnology: an untapped resource for food packaging. *Frontiers in Microbiology*, *8*, 1735.

Sharma, V. K., Siskova, K. M., Zboril, R., & Gardea-Torresdey, J. L. (2014). Organic-coated silver nanoparticles in biological and environmental conditions: fate, stability and toxicity. *Advances in Colloid and Interface Science*, *204*, 15–34.

Singh, T., Shukla, S., Kumar, P., Wahla, V., Bajpai, V. K., & Rather, I. A. (2017). Application of nanotechnology in food science: Perception and overview. *Frontiers in Microbiology*, *8*, 2517.

Sozer, N., & Kokini, J. L. (2009). Nanotechnology and its applications in the food sector. *Trends in Biotechnology*, *27*(2), 82–89.

Tager, J. (2014). Nanomaterials in food packaging: FSANZ fails consumers again. *Chain Reaction*, *122*, 16–17.

Tan, C., Feng, B., Zhang, X., Xia, W., & Xia, S. (2016). Biopolymer-coated liposomes by electrostatic adsorption of chitosan (chitosomes) as novel delivery systems for carotenoids. *Food Hydrocolloids*, *52*, 774–784.

Tharanathan, R. N. (2003). Biodegradable films and composite coatings: past, present and future. *Trends in Food Science & Technology*, *14*(3), 71–78.

Ubbink, J., & Krüger, J. (2006). Physical approaches for the delivery of active ingredients in foods. *Trends in Food Science & Technology*, *17*(5), 244–254.

Valdés, A., Mellinas, A. C., Ramos, M., Burgos, N., Jiménez, A., & Garrigós, M. D. C. (2015). Use of herbs, spices and their bioactive compounds in active food packaging. *RSC Advances*, *5*(50), 40324–40335.

Valencia, L., Nomena, E. M., Mathew, A. P., & Velikov, K. P. (2019). Biobased cellulose nanofibril–oil composite films for active edible barriers. *ACS Applied Materials & Interfaces*, *11*(17), 16040–16047.

Vieira, M. G. A., da Silva, M. A., dos Santos, L. O., & Beppu, M. M. (2011). Natural-based plasticizers and biopolymer films: A review. *European Polymer Journal*, *47*(3), 254–263.

Vilela, C., Kurek, M., Hayouka, Z., Röcker, B., Yildirim, S., Antunes, M. D. C., ... & Freire, C. S. (2018). A concise guide to active agents for active food packaging. *Trends in Food Science & Technology*, *80*, 212–222.

Wang, H., Du, L. J., Song, Z. M., & Chen, X. X. (2013). Progress in the characterization and safety evaluation of engineered inorganic nanomaterials in food. *Nanomedicine*, *8*(12), 2007–2025.

Wang, R., Xu, Y., Zhang, T., & Jiang, Y. (2015). Rapid and sensitive detection of Salmonella typhimurium using aptamer-conjugated carbon dots as fluorescence probe. *Analytical Methods*, *7*(5), 1701–1706.

Wang, Y., Zhang, R., Ahmed, S., Qin, W., & Liu, Y. (2019). Preparation and characterization of corn starch bio-active edible packaging films based on zein incorporated with orange-peel oil. *Antioxidants*, *8*(9), 391.

Weiss, J., Takhistov, P., & McClements, D. J. (2006). Functional materials in food nanotechnology. *Journal of Food Science*, *71*(9), R107-R116.

Wu, M., Tang, W., Guimar, J., Wang, Q., He, P., & Fang, Y. (2013). Electrochemical detection of Sudan I using a multi-walled carbon nanotube/chitosan composite modified glassy carbon electrode. *Am. J. Anal. Chem.*, *4*, 1–6.

Wu, S., Chen, X., Yi, M., Ge, J., Yin, G., Li, X., & He, M. (2019). Improving thermal, mechanical, and barrier properties of feather keratin/polyvinyl alcohol/tris (hydroxymethyl) aminomethane nanocomposite films by incorporating sodium montmorillonite and tio2. *Nanomaterials*, *9*(2), 298.

Wu, Y., Luo, Y., & Wang, Q. (2012). Antioxidant and antimicrobial properties of essential oils encapsulated in zein nanoparticles prepared by liquid–liquid dispersion method. *LWT-Food Science and Technology*, *48*(2), 283–290.

Yang, R., Zhou, Z., Sun, G., Gao, Y., Xu, J., Strappe, P., ... & Ding, X. (2015). Synthesis of homogeneous protein-stabilized rutin nanodispersions by reversible assembly of soybean (Glycine max) seed ferritin. *RSC Advances*, *5*(40), 31533–31540.

Yousuf, B., Qadri, O. S., & Srivastava, A. K. (2018). Recent developments in shelf-life extension of fresh-cut fruits and vegetables by application of different edible coatings: A review. *Lwt*, *89*, 198–209.

Yussuf, A. A., Al-Saleh, M. A., Al-Samhan, M. M., Al-Enezi, S. T., Al-Banna, A. H., & Abraham, G. (2018). Investigation of polypropylene-montmorillonite clay nanocomposite films containing a pro-degradant additive. *Journal of Polymers and the Environment*, *26*(1), 275–290.

Zambrano-Zaragoza, M. L., González-Reza, R., Mendoza-Muñoz, N., Miranda-Linares, V., Bernal-Couoh, T. F., Mendoza-Elvira, S., & Quintanar-Guerrero, D. (2018). Nanosystems in edible coatings: A novel strategy for food preservation. *International Journal of Molecular Sciences*, *19*(3), 705.

Zhang, H., Ma, X., Liu, Y., Duan, N., Wu, S., Wang, Z., & Xu, B. (2015). Gold nanoparticles enhanced SERS aptasensor for the simultaneous detection of Salmonella typhimurium and Staphylococcus aureus. *Biosensors and Bioelectronics*, *74*, 872–877.

Zhou, Y., Pan, F. G., Li, Y. S., Zhang, Y. Y., Zhang, J. H., Lu, S. Y., ... & Liu, Z. S. (2009). Colloidal gold probe-based immunochromatographic assay for the rapid detection of brevetoxins in fishery product samples. *Biosensors and Bioelectronics*, *24*(8), 2744–2747.

# Index

## A

abiotic stresses 105
acrylamide 20
active packaging 121–8, 130–3
adulteration 110–11, 116–17
agricultural production 104
agri-food business 98
amino acids 66, 69
anticaking agents 126
antimicrobial active packaging 121, 132
antimicrobial films 121–8, 132
antioxidants 69
arsenic 19

## B

bacteriocins 52
behavioral intentions 36, 41
bioactives 66–9
bio-based packaging 121–9
bio-degradability 3
bioengineering 46, 52–3, 55
biofilms 52
biofuels 105
biomass 87
bionanoencapsulated 53
bionanosensors 123, 134–5
braising 20
brewing 84

## C

cadmium 19
calorific content 62
canning 20
carbohydrate polymers 4
carbonylation 22
carvacrol 2
CCPS 25, 28–9
chilling stress 103
cohesive structures 3
cold climate 61
contamination 20–4, 29, 111–13, 115–16
coumarins 2
COVID-19 110
critical limit 25, 28
crop production 62, 99, 105
cross-contamination 20

## D

deceptive conduct 118–19
desiccation 83
deterioration 2, 48
diabetes 65
dielectric heating 47–50

dietary approaches 52
dietary pollutants 18
dietary shift 104
diversity 61
drought stress 95

## E

eco-friendly 34, 37, 38
eco-friendly restaurant operations 36
energy efficiency 50, 83
environmental protection agency 34–7
environmental degradation 34
environmental friendliness 112
environmentally friendly 36–8, 41–2
environmentally friendly procedures 38
epidemiological studies 64
ethylene vinyl alcohol (EVOH) 2
eugenol 2

## F

farming system 105
fatty foods 19
flavones 2
flavonoids 2
flavonols 2
food and drug administration 112
food-borne infections 111
food contamination 19–20
food hygiene 18, 24, 29, 30, 112
food industries 8, 24, 53, 80, 89, 90, 96, 112
food intoxication 109, 115
food losses 99
food matrix 64
food packaging 121–3, 127–9, 131–7
food poisoning 109, 115
food pollutants 18
food processing 18–21
food quality 109–13
food safety 60, 63, 82, 109–20
food safety and standards authority of India 112–13, 117
food safety issues 18, 63, 110, 121, 135
food storage 112
food supply chain 20
fossil energy requirements 39
fraudulently advertises 119
freezing stress 103
frying 20
fungicides 20

## G

gastrointestinal microbiota 18
gelling agents 121–4, 126–7
global food security 96, 97, 101, 103, 105

globalization 113
glucose 21
glycemic index 66
glycerol 21
good handling practices (GHP) 116
good manufacturing practices (GMP) 23, 116
green hotel 36
green image 38, 41–2
green marketing 42
green practice 33–42
green restaurant association 34

## H

HACCP 23–9
hazard 20, 22–9
hazard analysis 23–5, 27–8
hazard analysis critical control points (HACCP) 116
heat retrieval 84
heavy metal 19, 104
high hydrostatic pressure 45, 47–8
homogenization 49
hospitality business 35
hybrid heating 85

## I

improved packaging 71, 128, 130
insecticides 19

## L

lead 19
life cycle assessment 39
lifestyle habits 62
lipid-hydroperoxide 21

## M

macroscopic 2
maillard reaction 21
malathion 22
malignant cells 22
mellowness 22
mercury 19
metabolic syndrome 62, 65
metal stress 104
microbes 47, 52, 69
microorganisms 22–3
microscopic 2
microstructures 63
monetary 89
mutagens 21
mycotoxins 22

## N

nanoadditives 127
nano-carriers 121–4

nanoencapsulation 121–7, 132
nanotechnology 47, 53, 70, 121–8, 130, 134–5
nitrogen components 20
nitrosamines 20
nitrosodimethylamine 20
non-caloric processing 86
non-potable 35
novel cooling cycles 85
nutraceuticals 123
nutrients 60–9, 71
nutritional 110–11
nutritional value of food 122

## O

octanol-water 22
off-coloration 20
off-flavors 20
ohmic heating 86
optimization 49, 87
organic food 41
organic pollutants 19
organochlorine 19
osmotic stress 102
oxidation 20
oxygen indicators 121, 135
oxygen scavenging films 121, 133

## P

packaging 18–22
packaging and labelling laws 117
parching 84
pathogen 114–17
perceived customer effectiveness 37
phenolic acids 2
phenols 2
photovoltaic cells 88
phthalate esters 21
plasticizers 21
polyamide (PA) 2
polycyclic aromatic hydrocarbons (PAHS) 20
polyethylene (PE) 2
polyethylene terephthalate (PET) 2
polymers 2–10, 21, 64, 71
polystyrene (PS) 2
polyvinylchloride (PVC) 2
pre-packaged food 118
preservation 47–9, 51
preserve 2, 50–3, 62–3, 81
pressurization 47
proprietary food 117–18
PTSC system 88
pullet protein 22
pulsed light 51

## Q

quinines 2

# INDEX

## R

recyclable 40-2
recycle 37, 41
roasting 20

## S

salinity stress 102
salmonella 113, 115
sanitary standards 119
secondary metabolites 2
sequential ventilation 87
shelf life 52-3, 63, 81, 86, 96
slack resources 38
smart/ intelligent packaging 121, 134
social responsibility 40
social sensitivity 40
SSOP 23-4
stabilizing agents 111
stockpiling 84
storage 18-23
styrofoam 41
sucrose 67
sustainability initiatives 35
sustainable development 120
sustainable food processing 33-9
sustainable food sources 33-9
sustainable food systems 33-9

synthetic biology 47, 52
synthetic biology 47, 52

## T

tannins 2
temperature stress 103
texture modifiers 121
thermal conductivity 50
thermal stability 2, 22
thermal treatment 46, 60
thymol 2
tocopherol 21
topographical 67
toxic 19, 22, 60, 64
toxicants 18
transportation 19-20, 22-9
trash management 38

## U

ultrasonication 49
UV absorbing films 121, 133

## W

waste management 63
whole grain 65
wholesome food 117

Made in the USA
Monee, IL
03 May 2026